FATIH AKAY

Chemistry for the Curious

This book is dedicated to those who seek to understand the fundamental principles of chemistry and the world around us. May the knowledge within these pages inspire you to ask questions and seek answers, and may it encourage you to appreciate the beauty and complexity of the chemical world.

Thank you for joining me on this journey of exploration.

"The universe is full of magical things patiently waiting for our wits to grow sharper."

– Eden Phillpotts

Contents

Preface

Dear Reader,

I am thrilled to present to you my latest book that explores some of the fascinating mysteries of the world of science. Science has always been a source of wonder for me, and it has played a vital role in shaping our understanding of the world we live in. The beauty of science lies in its ability to reveal the hidden secrets of the universe, unraveling the mysteries of nature, and providing us with answers to some of the most intriguing questions.

In this book, I have delved into various topics ranging from the chemistry of everyday objects to the principles that govern the behavior of our planet. The aim of this book is to take you on an exciting journey of discovery, where you will learn about the inner workings of our world in a way that is both informative and engaging.

I have tried to write this book in a way that is accessible to everyone, regardless of their background or level of scientific knowledge. My hope is that by the end of this book, you will not only have a deeper understanding of the world around us but also an appreciation for the remarkable discoveries that have been made over the centuries.

Science is a never-ending journey of exploration and discovery, and in this book, I hope to spark your curiosity and inspire you to learn more about the world around us. My goal is to make science accessible and engaging, to show you how science impacts our lives every day, and to encourage you to ask questions and seek out answers.

This book is not intended to be a comprehensive guide to science but rather a collection of fascinating insights into various topics. I hope that after reading this book, you will come away with a greater appreciation for the natural world and the role that science plays in helping us understand it.

I would like to thank you for taking the time to read this book, and I hope that you will find it as enjoyable to read as it was for me to write. Whether you are a seasoned scientist or simply someone who is curious about the world, I believe that this book has something to offer everyone.

Thank you for joining me on this journey of exploration and discovery. Let's dive in!

So sit back, relax, and let us embark on this journey of discovery together.

Sincerely,

Acknowledgement

I would like to express my sincere gratitude to everyone who has contributed to the creation of this book.

First and foremost, I would like to thank my family and friends for their unwavering support throughout this journey. Their encouragement and belief in me has been a constant source of motivation.

I would like to extend my heartfelt appreciation to my mentor, who has been an invaluable guide and a source of inspiration. Their guidance and encouragement have been instrumental in shaping my thoughts and ideas.

I would also like to thank all the experts and professionals who have generously shared their knowledge and expertise with me. Their insights have added immense value to this book.

Lastly, I would like to express my gratitude to my publisher, editors, and everyone involved in the production of this book. Their dedication and hard work have brought this project to fruition.

To all those who have contributed in some way, I offer my heartfelt thanks. This book would not have been possible without your support and encouragement.

I

Part One

1

Why is it the most stable state to have 8 electrons in the outermost orbit of atoms?

All atoms have different numbers of electrons in a neutral state, and these electrons are arranged in energy levels called periods around the nucleus. The maximum number of electrons that each period can hold is determined by the 2n2 rule (n is the period number), which is 2, 8, 18, 32, etc. Except for hydrogen (1H), helium (2He), lithium (3Li), beryllium (4Be), and boron (5B), the most stable state for all other elements is to have 8 electrons in their outermost shell. This is because it is easier for these elements to reach 2 than to reach 8.

The stability of the electron shells in each period increases as the orbitals become more filled. In each period, the s orbital can hold 2 electrons, the p orbital can hold 6 electrons, the d orbital can hold 10 electrons, and the f orbital can hold 14 electrons. There is a difference in energy levels between these orbitals in each period, with the largest energy difference occurring between the p and d orbitals. This means that it is difficult for an atom that has filled its p orbital to also fill its d orbital.

An atom that has filled its p orbital already has 8 electrons, due to the fact that its s orbital was filled before. As much as possible, this structure is maintained to prevent destabilization.

2

How do clouds form?

Clouds are one of the most beautiful natural phenomena, but have you ever wondered how they are formed? Cloud formation is a complex process that involves the movement of water molecules in the atmosphere.

Water is in constant motion in the oceans, lakes, and rivers. As we know, water evaporates between 0°C to 100°C, and as the temperature increases, the rate of evaporation also increases. The density of the water vapor is lower than that of the surrounding air, causing the water vapor to rise upwards.

When water molecules in the gas state rise higher into the atmosphere, they encounter cold air masses. This cold air lowers the temperature, causing the water vapor to condense into liquid droplets. Aerosols, which are tiny particles of dust, dirt, and other substances floating in the air, act as nuclei for the water droplets to form around.

The process of condensation continues as more water molecules accumulate around the aerosols, forming a cloud. The size of the cloud depends on the amount of moisture in the air and the number of aerosol particles present. Different types of clouds are formed depending on the altitude, temperature, and humidity of the air.

Clouds play an essential role in regulating Earth's climate by reflecting and absorbing the sun's radiation. They also have a significant impact on the water cycle by releasing rain, snow, and hail. Understanding the process of cloud formation is crucial for meteorologists and climate scientists to make accurate weather predictions and study the Earth's climate.

In conclusion, clouds are formed through the condensation of water vapor in the atmosphere around aerosol particles. The process of cloud formation is fascinating and plays a critical role in the Earth's climate and water cycle.

3

Is the reason why mercury (Hg) is liquid rare?

Mercury (Hg) is a unique element in many ways, one of which is its liquid state at room temperature. Unlike most metals, which are solid at room temperature, mercury is a liquid metal. This interesting characteristic of mercury can be explained by its bonding behavior.

In the case of mercury, the 6s orbital is completely filled, and the contribution of metallic bonding is minimal. The bonds between mercury atoms are Van der Waals type, which are relatively weak compared to metallic bonds. Since mercury is a heavy element and the Van der Waals bond is strong, mercury cannot easily transition to the gas phase and remains in a liquid state at normal conditions.

This low boiling point of mercury, combined with its conductivity and high density, makes it a valuable metal in many industrial and scientific applications. It is commonly used in thermometers, barometers, electrical switches, and dental fillings, among other things.

However, it is important to handle mercury with caution as it is toxic to humans and the environment. When heated or spilled, mercury can release harmful vapors that can cause serious health problems. Thus, it is crucial to handle it safely and dispose of it properly.

In conclusion, the liquid state of mercury at room temperature is due to its unique bonding behavior, and its properties make it a useful metal in many applications. However, it is important to handle it with care to avoid any potential harm.

4

Why do only a few metals respond to magnets?

Metals such as iron, cobalt, nickel, gadolinium, and neodymium are known to be magnetic, while many other metals are not. The reason for this lies in the electronic structure of these metals. Metals such as iron, nickel, and chromium have partially-filled d orbitals, which contain unpaired electrons. These unpaired electrons possess magnetic moments and can align themselves in a parallel manner due to a phenomenon known as "long range ordering." This alignment results in the creation of magnetic domains, which are regions of the metal where the magnetic moments are parallel to each other, resulting in a high magnetic field.

When an external magnetic field is applied, it interacts with these magnetic domains, causing them to align themselves in the same direction as the external field. This results in a net magnetic force, which is why these metals are attracted to magnets. This phenomenon is known as ferromagnetism.

Interestingly, not all magnetic metals are ferromagnetic. Other types of magnetism include paramagnetism and diamagnetism, which are exhibited by a wider range of materials. However, the ferromagnetic metals such as iron, cobalt, nickel, gadolinium, and neodymium are the most well-known due to their strong magnetic properties and practical applications in various industries.

In summary, the reason why only a few metals are attracted to magnets lies in their electronic structure and the resulting creation of magnetic domains. Metals with partially-filled d orbitals and unpaired electrons are able to exhibit ferromagnetism due to the parallel alignment of their magnetic moments. This phenomenon results in a net magnetic force, which allows these metals to be attracted to magnets.

5

What does it mean when a battery runs out, and how does a rechargeable battery work?

Batteries are an essential part of modern life, powering everything from our phones and laptops to our cars and homes. However, all batteries have a limited lifespan, and eventually, they will run out of power. When this happens, we say that the battery has "died" or "run out." In technical terms, this means that the electrochemical reaction inside the battery has come to a halt, usually because one of the reactants has been depleted.

Rechargeable batteries, also known as secondary batteries, work by reversing the electrochemical reaction that powers them. Instead of producing electricity, they consume it, using an external power source to drive the reaction in the opposite direction. This is how they can be charged and recharged multiple times, as long as the reactants remain available.

The most common type of rechargeable battery is the lithium-ion battery, which is widely used in portable electronic devices like smartphones, laptops, and tablets. The basic principle behind a lithium-ion battery is the movement of lithium ions between the two electrodes, typically made of graphite and a metal oxide, during charging and discharging.

When the battery is charged, lithium ions are driven from the metal oxide cathode to the graphite anode, where they are stored as lithium atoms. During discharge, the process is reversed, with lithium ions moving back to the cathode and releasing their stored energy in the form of electrical current.

Over time, the performance of a rechargeable battery will degrade due to a number of factors, including the loss of active material from the electrodes, the buildup of unwanted compounds like lithium plating, and the corrosion of internal components. Eventually, the battery will no longer be able to hold a charge, and it will need to be replaced.

In conclusion, when a battery runs out of power, it means that the electrochemical reaction inside the battery has ceased, usually due to the depletion of one of the reactants. Rechargeable batteries work by reversing this reaction, using an external power source to recharge the battery and restore its energy storage capacity. While rechargeable batteries can be used and recharged multiple times, they will eventually degrade and need to be replaced.

6

How do you explain passing through clouds when the outside temperature on an airplane is -20 to -30 degrees Celsius?

When flying in an airplane, it's common to experience temperatures below freezing, even when passing through clouds. This may seem counterintuitive since the outside temperature at cruising altitude can often reach as low as –20 to –30 degrees Celsius. However, there is a scientific explanation behind this phenomenon.

The atmosphere is divided into different layers, with the troposphere being the lowest and most dense layer where all weather conditions occur. As the altitude increases, the pressure and temperature decrease, and the air becomes thinner. This is why it becomes colder at higher altitudes.

Moreover, the air in the upper atmosphere contains very little moisture or water vapor. When the airplane passes through clouds, the tiny droplets of water or ice crystals present in the clouds are much colder than the surrounding air, and they freeze instantly on contact with the airplane's wings, fuselage, and engines. This process is known as icing, and it can be hazardous if not managed correctly.

However, the presence of aerosols in the atmosphere can alter the freezing point of the moisture in the air. Aerosols are tiny particles suspended in the air that can originate from natural sources such as dust, volcanic eruptions, and sea salt or human-made sources such as pollution. These particles act as nuclei around which the water droplets or ice crystals can form, and they can affect the freezing point of the moisture.

In the upper atmosphere, where the air is very thin, the concentration of aerosols is higher. This increased concentration of aerosols lowers the freezing point of the moisture, allowing clouds to exist at much lower temperatures than they would otherwise. As a result, when an airplane flies through these clouds, the moisture freezes on contact, despite the outside air temperature being -20 to -30 degrees Celsius.

In conclusion, the presence of aerosols in the upper atmosphere is what allows clouds to exist at extremely low temperatures. This is why when flying in an airplane, even when the outside temperature is very cold, passing through clouds can result in ice forming on the aircraft. It's essential for pilots to be aware of this phenomenon and take appropriate measures to prevent icing and ensure a safe flight.

7

Which metal melts when we hold it in our hand?

Metals are generally solid, whether they are hard or soft. However, there are some exceptions to this rule, and some metals are more liquid-like. There are certain metals that can even melt in your hand when you hold them. Gallium and cesium are two examples of such metals, as their melting points are just slightly above 30 degrees Celsius. Francium is another metal that can melt at room temperature, but it is not yet readily available in pure form.

Mercury is another metal that has unique properties. It is a liquid at room temperature, with a freezing point of -39 degrees Celsius and a boiling point of 300 degrees Celsius. This makes it an excellent material for use in various types of thermometers. However, due to its low boiling point, it cannot be used to measure high temperatures.

Gallium, on the other hand, has a boiling point of approximately 2,000 degrees Celsius, making it an excellent material for high-temperature thermometers. It is also a good alternative to mercury in some applications, such as in certain types of switches.

Some metals have incredibly high boiling points. For example, hafnium begins to boil at a temperature of approximately 5,400 degrees Celsius, which is roughly equivalent to the temperature on the surface of the sun. This makes hafnium an excellent material for use in high-temperature applications, such as in nuclear reactors.

In conclusion, while most metals are solid at room temperature, there are some exceptions. Some metals, like gallium and cesium, can even melt in your hand due to their low melting points. Mercury is unique in that it is a liquid at room temperature, while other metals like hafnium have incredibly high boiling points, making them ideal for high-temperature applications. The properties of each metal determine its suitability for different applications, and understanding these properties is important in selecting the right material for a given use.

8

Why is the boiling point of arsenic lower than its melting point?

Arsenic, or As, is a chemical element with the atomic number 33. It is classified as a metalloid and is commonly found in minerals such as realgar, orpiment, and arsenopyrite. One interesting characteristic of arsenic is that its boiling point is lower than its melting point.

At standard temperature and pressure (STP), which is defined as 0°C (273.15 K) and 1 atmosphere (atm) of pressure, arsenic is a solid. However, if it is heated, it will sublime directly from a solid to a gas, bypassing the liquid phase. This means that its boiling point, which is the temperature at which it changes from a liquid to a gas, is actually lower than its melting point, which is the temperature at which it changes from a solid to a liquid.

The boiling point of arsenic is approximately 614°C, while its melting point is 817°C. This is due to the unique electronic structure of arsenic atoms, which have five valence electrons in their outermost shell. This electronic configuration causes arsenic to form covalent bonds in a molecular lattice structure, rather than metallic bonds as found in most metals.

As a result, arsenic has a relatively weak atomic bonding, which causes the molecules to have low intermolecular forces of attraction. This means that when heated, the arsenic molecules will easily break free from the lattice and transition to the gas phase, without first passing through a liquid phase.

It is worth noting that while arsenic has a lower boiling point than its melting point, this is not true for all elements. Most elements have a higher boiling point than their melting point, as the increased thermal energy required to break atomic bonds and transition from a solid to a gas phase is usually greater than that required to transition from a solid to a liquid phase.

In conclusion, the unique electronic structure of arsenic atoms leads to a lower boiling point than its melting point. When heated, arsenic molecules will easily transition from a solid to a gas phase, bypassing the liquid phase. Understanding the properties and behavior of elements like arsenic can help us better understand the chemical and physical world around us.

9

Do the weights of atoms change?

The weights of atoms do not change, but the way we measure them does.

When calculating atomic weights, a specific atom is chosen as a reference point, and the atomic mass unit (amu) is calculated based on that atom. Initially, the atomic weights were measured relative to the hydrogen (H) atom, but later calculations were based on the oxygen (O) atom. Nowadays, the atomic weight of other atoms is calculated based on the 12C atom as a reference point, since it is found naturally in nature in a percentage of approximately 99%. It has been shown that this calculation method is more accurate.

However, when calculating the average atomic weight, radioactive isotopes are not considered. The calculation of atomic weights has changed over time as different reference atoms have been used. Although the quantity of atoms has not changed, the numerical values used to describe atomic weights have changed.

In conclusion, the weights of atoms themselves do not change, but the way we measure them does. As science advances and technology improves, the methods we use to measure atomic weights may continue to evolve, but the fundamental principles of atomic structure and weight will remain constant.

10

How can you tell real honey from fake honey?

The consumption of honey has been a part of human culture for thousands of years. Honey is a natural sweetener that is rich in antioxidants, vitamins, and minerals. However, with the rise in commercial production, there is an increase in the production of adulterated honey, which has led to consumers being confused about the authenticity of the honey they purchase. So, how can you tell if your honey is real?

One way to determine the authenticity of honey is to perform a flame test. Real honey contains glucose, which burns easily. To conduct the flame test, dip a matchstick in honey and strike it against the matchbox. If the honey is real, the matchstick will light easily, and the flame will burn steadily. However, if the honey is fake, it will not catch fire because it contains water, corn syrup, or other additives that prevent it from burning.

Another method to detect real honey is to conduct a water test. Mix a tablespoon of honey into a glass of water and observe. Real honey will settle at the bottom of the glass, while fake honey will dissolve quickly, forming a cloudy solution.

It is also essential to check the label of the honey jar. Real honey should not have any added ingredients like high fructose corn syrup or other artificial sweeteners. Check for the honey's country of origin, and if it is sourced from one country, it is more likely to be authentic. However, honey that is sourced from multiple countries or that is labeled as "blended" may not be genuine.

Lastly, it is crucial to purchase honey from a reputable source. Local beekeepers and farmers' markets are an excellent source for real honey, as they are more likely to be selling raw, unprocessed honey that has not been adulterated.

In conclusion, it is possible to determine the authenticity of honey through various tests and checks. Real honey should be free of added ingredients, settle at the bottom of a water glass, and catch fire easily when exposed to flame. By being aware of these methods, consumers can be confident that they are purchasing real honey and supporting local beekeepers and farmers.

11

When we say that everything was once in the form of a gas cloud, what gas are we referring to?

It is a widely accepted scientific theory that the universe and everything in it, including our solar system, originated from a gas cloud. This gas cloud is believed to have been mostly made up of hydrogen and helium, but other elements and compounds were also present. One of these compounds is methane, which is the gas that we believe played a significant role in the formation of our solar system.

Methane is a simple compound made up of one carbon atom and four hydrogen atoms. It is a colorless, odorless gas that is the main component of natural gas. Methane is found on Earth in a variety of sources, including fossil fuels, wetlands, and the digestive systems of some animals.

The theory is that billions of years ago, a cloud of gas and dust in our galaxy began to collapse under its own gravity. As it collapsed, it began to spin faster, and as it spun, it flattened into a disk. The center of this disk became very hot and dense, eventually becoming the sun. The rest of the disk was made up of gas, dust, and small rocky and icy particles called planetesimals.

As these planetesimals collided and stuck together, they gradually grew larger, eventually becoming the planets in our solar system. Methane is believed to have played a key role in this process by acting as a kind of glue that held together the dust and ice particles as they collided and formed larger and larger bodies.

So, when we say that everything was once a gas cloud, we are referring to a cloud of mostly hydrogen and helium, with some other elements and compounds, including methane. This gas cloud is thought to have collapsed and formed our solar system, with methane playing an important role in the process.

12

What is the chemical composition of gasoline, which is used as fuel for cars?

Gasoline, also known as petrol, is a widely-used fuel for internal combustion engines, such as those found in cars and motorcycles. It is a complex mixture of hydrocarbons, which are compounds made up of carbon and hydrogen atoms. The exact composition of gasoline can vary depending on the source of crude oil and the refining process used to create it, but in general, gasoline is composed mainly of alkanes, cycloalkanes, and aromatic hydrocarbons.

Alkanes, also known as paraffins, are straight-chain or branched hydrocarbons with only single bonds between the carbon atoms. The most common alkanes found in gasoline are pentane, hexane, and heptane. These alkanes have low boiling points, making them volatile and easily vaporized. This property is important for gasoline because it allows it to evaporate easily and mix with air to form a combustible mixture.

Cycloalkanes, also known as naphthenes, are hydrocarbons that contain one or more rings of carbon atoms. They are similar to alkanes in that they only have single bonds between the carbon atoms, but the ring structure of cycloalkanes makes them more stable than their straight-chain counterparts. Cycloalkanes found in gasoline include cyclohexane and methylcyclopentane.

Aromatic hydrocarbons, such as benzene, toluene, and xylene, are compounds that contain one or more rings of carbon atoms with alternating double bonds. These double bonds create a unique aromatic smell, which is where the name "aromatic" comes from. Aromatic hydrocarbons are more stable than alkanes and cycloalkanes, but they also have higher boiling points, making them less volatile. Aromatic hydrocarbons are added to gasoline to increase its octane rating, which measures the fuel's resistance to knocking or pinging in an engine.

In addition to these three main components, gasoline can also contain small amounts of oxygen-containing compounds, such as alcohols and ethers, as well as trace amounts of sulfur, nitrogen, and other elements. These impurities can have negative effects on engine performance and contribute to air pollution, so regulations limit the amount of these compounds that can be present in gasoline.

The composition of gasoline can vary depending on the type of crude oil used as the starting material and the refining process used to create the final product. Crude oil is a mixture of hydrocarbons of various lengths and structures, and the refining process separates these compounds based on their boiling points. This process, known as fractional distillation, involves heating the crude oil and collecting the vaporized hydrocarbons at different temperatures. The resulting fractions are then further processed to remove impurities and adjust the properties of the final product.

One important property of gasoline is its octane rating, which measures the fuel's resistance to knocking or pinging in an engine. Knocking occurs when the fuel mixture in the engine ignites spontaneously and out of sequence, causing a sharp metallic noise and potentially damaging the engine. Higher-octane gasoline is more resistant to knocking, allowing engines to run at higher compression ratios and produce more power.

To increase the octane rating of gasoline, manufacturers can add certain additives, such as aromatics, oxygenates, and metal-containing compounds. These additives can improve combustion efficiency, reduce emissions, and enhance engine performance. However, they can also have negative effects on air quality and human health, so regulations limit the amount and type of additives that can be used in gasoline.

In recent years, there has been growing interest in alternative fuels to replace gasoline, such as biofuels, electric vehicles, and hydrogen fuel cells. Biofuels, such as ethanol and biodiesel.

13

What is natural gas?

Natural gas is a type of fossil fuel that is primarily composed of methane, along with small amounts of other hydrocarbons, such as ethane, propane, and butane. It is a popular fuel source for heating, cooking, and generating electricity, and is considered to be one of the cleanest burning fossil fuels. In this article, we will explore the chemical composition of natural gas, how it is formed, its uses, and its impact on the environment.

Chemical Composition of Natural Gas

As previously mentioned, natural gas is primarily composed of methane, which is a colorless, odorless gas that is highly flammable. Methane molecules are made up of one carbon atom and four hydrogen atoms, giving it the chemical formula CH_4. In addition to methane, natural gas can also contain small amounts of other hydrocarbons, such as ethane, propane, and butane.

The exact composition of natural gas can vary depending on where it is sourced from. For example, natural gas sourced from shale formations, known as shale gas, tends to have a higher proportion of ethane and propane compared to natural gas sourced from conventional gas reservoirs. Additionally, natural gas can also contain small amounts of impurities, such as water vapor, nitrogen, and carbon dioxide.

Formation of Natural Gas

Natural gas is formed from the remains of dead plants and animals that have been buried and subjected to high pressure and temperature over millions of years. As these organic materials are buried deeper and deeper into the Earth's crust, the high temperature and pressure cause the organic matter to break down into hydrocarbons, which eventually migrate towards the surface of the Earth.

The hydrocarbons can accumulate in porous rock formations, such as sandstone and limestone, and form natural gas reservoirs. Natural gas can also be found in association with oil deposits, as the two are often formed from the same source rocks and can coexist in the same reservoir.

Uses of Natural Gas

Natural gas is a versatile fuel source that is used for a wide range of applications. One of the most common uses of natural gas is for heating and cooking in homes and businesses. Natural gas furnaces and water heaters are popular choices for homeowners, as they are efficient and relatively inexpensive compared to other heating sources.

Natural gas is also used for generating electricity, either by burning it in a gas turbine or by using it to fuel a steam turbine. This process, known as natural gas-fired power generation, is becoming increasingly popular due to the lower emissions compared to other fossil fuels, such as coal.

In addition to its uses in heating, cooking, and electricity generation, natural gas is also used as a feedstock for the production of chemicals and other materials. For example, ethane, a component of natural gas, can be used as a raw material for the production of ethylene, which is used to make a wide range of plastics and synthetic materials.

Environmental Impact of Natural Gas

While natural gas is considered to be one of the cleanest burning fossil fuels, its extraction and use still have an impact on the environment. One of the main environmental concerns associated with natural gas is the release of methane, a potent greenhouse gas, during production and transportation.

Methane can leak from natural gas wells, pipelines, and storage facilities, and has a much higher global warming potential than carbon dioxide. In addition, the process of hydraulic fracturing, or fracking, which is used to extract shale gas from underground formations, has been linked to water contamination and other environmental issues.

Another environmental concern associated with natural gas is the impact on local ecosystems and habitats. Natural gas drilling and production can disrupt wildlife habitats and cause soil erosion, while the construction of pipelines and other infrastructure can fragment forests and other natural areas.

14

What percentage of our body is made up of C, O, and H atoms?

Carbon (C), oxygen (O), and hydrogen (H) are three of the most abundant elements in the human body, making up the vast majority of our chemical composition.

Carbon is a key component of many of the molecules in our body, including proteins, fats, and DNA. It is also essential for energy production, as it is a key element in glucose, which fuels our cells. The human body is about 18.5% carbon by mass, making it the fourth most abundant element in our bodies after oxygen, hydrogen, and nitrogen.

Oxygen is the most abundant element in the human body, accounting for about 65% of our total mass. It is a key component of water and is required for cellular respiration, which is the process by which our cells produce energy. Oxygen is also found in many other molecules in our bodies, including proteins, nucleic acids, and lipids.

Hydrogen is the third most abundant element in the human body, making up about 9.5% of our total mass. It is a key component of water and is also found in many organic molecules, including proteins, carbohydrates, and lipids. Hydrogen is also a key player in cellular respiration, where it combines with oxygen to produce water and energy.

Other elements found in the human body include nitrogen (3.3%), calcium

(1.5%), phosphorus (1.0%), and potassium (0.25%). These elements play important roles in many of the body's functions, including muscle and nerve function, bone health, and energy production.

Nitrogen is a key component of amino acids, which are the building blocks of proteins. It is also found in nucleic acids, which are the molecules that make up our DNA. Calcium is essential for strong bones and teeth, as well as proper muscle function and blood clotting. Phosphorus is also important for bone and teeth health, as well as energy production and DNA synthesis. Potassium is critical for nerve and muscle function, and also helps regulate blood pressure and fluid balance in the body.

Other trace elements found in the human body include iron, zinc, copper, iodine, and selenium. These elements are required in very small amounts but play important roles in various body functions, such as the production of red blood cells (iron) and the functioning of enzymes (zinc and copper).

In summary, while carbon, oxygen, and hydrogen are the most abundant elements in the human body, many other elements also play crucial roles in our overall health and well-being. Maintaining a balance of these elements through a healthy diet and proper nutrition is essential for optimal health.

15

What are the alkali metals found in soil?

Alkali metals are a group of chemical elements with similar properties that are found in the first column of the periodic table. These elements are known for their reactivity, high conductivity, low melting and boiling points, and ability to form basic oxides. Alkali metals are essential for many industrial and technological applications, but they are also important in nature. In this article, we will explore the alkali metals found in soil and their role in the environment.

The alkali metals found in soil are primarily sodium, potassium, and lithium. These elements are essential for plant growth and are taken up by the roots of plants from the soil. Sodium is particularly important for maintaining the osmotic balance of plant cells and is necessary for the proper functioning of enzymes. Potassium is also essential for enzyme function and is involved in regulating the water balance of plants. Lithium, although not as well-studied as sodium and potassium, has been shown to promote plant growth and development.

Sodium is the most abundant alkali metal found in soil. It is present in soil minerals and is released into the soil solution through weathering processes. Sodium is typically found in higher concentrations in arid and semiarid regions where the rate of weathering is low and evapotranspiration is high. The amount of sodium in soil can also be influenced by human activities such as irrigation with saline water, fertilization with sodium-containing fertilizers, and the use of sodium-based deicing salts.

Potassium is the second most abundant alkali metal found in soil. It is also present in soil minerals and is released into the soil solution through weathering processes. Potassium is typically found in higher concentrations in soils that are rich in clay minerals or have high organic matter content. The amount of potassium in soil can be influenced by human activities such as fertilization with potassium-containing fertilizers and the use of potassium-based deicing salts.

Lithium is the least abundant of the alkali metals found in soil. It is present in soil minerals such as micas and feldspars and is released into the soil solution through weathering processes. Lithium is typically found in very low concentrations in soil, but it has been shown to promote plant growth and development at low concentrations.

The concentration of alkali metals in soil can vary widely depending on a variety of factors such as climate, soil type, and land use. In general, soils in arid and semiarid regions tend to have higher concentrations of alkali metals due to the low rate of weathering and high evapotranspiration. Soils that are rich in clay minerals or have high organic matter content also tend to have higher concentrations of alkali metals.

While alkali metals are essential for plant growth, they can also have negative effects on the environment if present in high concentrations. Excess sodium in soil can lead to soil salinization, which can reduce plant growth and yield. High concentrations of sodium in groundwater can also lead to soil and water contamination. Excess potassium in soil can lead to nutrient imbalances and can also contribute to soil salinization. High concentrations of lithium in soil have not been shown to have negative effects on the environment.

In addition to their role in plant growth, alkali metals are also important in many industrial and technological applications. Sodium is used in the production of glass, soap, and other chemicals. Potassium is used in the production of fertilizers, detergents, and other chemicals. Lithium is used in the production of batteries, ceramics, and other materials.

In recent years, there has been growing interest in the use of alkali metals as energy storage materials. Lithium-ion batteries, which are widely used in portable electronic devices, use lithium as the active material in the battery electrode. Sodium-ion batteries, which are still under development, have been proposed as a promising alternative to lithium-ion batteries. Sodium is a highly abundant element and can be sourced more sustainably and inexpensively than lithium. In addition, sodium-ion batteries have the potential for high energy density and fast charging rates.

Like lithium-ion batteries, sodium-ion batteries work by shuttling ions between a positive and negative electrode. However, the differences in the size and charge of sodium ions versus lithium ions require different materials for the electrodes.

The positive electrode, or cathode, of a sodium-ion battery typically consists of a material that can intercalate (or insert) sodium ions, such as a sodium transition metal oxide. The negative electrode, or anode, is typically made of carbon or graphite, which can intercalate sodium ions as well.

One challenge with sodium-ion batteries is their lower energy density compared to lithium-ion batteries. This means that sodium-ion batteries may not last as long on a single charge and may require more frequent charging. In addition, the materials used for sodium-ion batteries are still being developed and optimized for better performance.

Despite these challenges, sodium-ion batteries hold promise for a range of applications, from large-scale energy storage systems to powering electric vehicles. Their abundance and lower cost make them an attractive alternative to lithium-ion batteries, especially for regions with limited access to lithium resources.

Research into sodium-ion batteries is ongoing, with scientists working to improve their performance and durability. As with any emerging technology, there are still many obstacles to overcome before sodium-ion batteries can be widely adopted, but the potential benefits make it an exciting field of research to watch.

16

How does thunder occur

Thunder is a natural phenomenon that accompanies lightning, a discharge of electricity between clouds or between a cloud and the ground. Thunder is the sound that results from this electrical discharge, and it is often described as a loud, rumbling noise that can shake the ground and rattle windows.

So how exactly does thunder occur? The process begins with the formation of a thunderstorm, which typically happens when warm, moist air rises and cools, forming clouds. Inside these clouds, water droplets and ice particles collide and create electrical charges. Positive charges build up near the top of the cloud, while negative charges accumulate near the bottom.

As the storm intensifies, the negative charges at the bottom of the cloud begin to attract positive charges on the ground. When the electrical potential between the cloud and the ground becomes large enough, a lightning bolt can occur. Lightning is essentially a giant spark of electricity that ionizes the air, creating a channel of plasma through which the electric current flows.

The intense heat of the lightning bolt, which can reach temperatures of up to 30,000 kelvins (53,540 degrees Fahrenheit), causes the air to rapidly expand and contract. This creates a shock wave that travels through the air, which we hear as thunder.

The sound of thunder is actually a sonic boom, similar to the sound of a supersonic jet breaking the sound barrier. As the shock wave travels through the air, it compresses the air in front of it and creates a region of high pressure.

This high-pressure wave moves through the air, causing the rapid vibrations that we hear as thunder.

The distance between lightning and thunder can be used to estimate how far away the lightning struck. Sound travels through the air at a speed of approximately 1,125 feet (340 meters) per second, while light travels at a much faster speed of approximately 186,000 miles (299,792 kilometers) per second. By counting the number of seconds between a flash of lightning and the sound of thunder, you can estimate the distance to the lightning strike in miles or kilometers. For example, if you hear thunder 10 seconds after seeing a lightning flash, the lightning is approximately 2 miles (3 kilometers) away.

In addition to the sound of thunder, lightning can also cause other atmospheric phenomena. For example, it can create a variety of visible light effects, such as a bright flash of light, a glowing cloud, or even a fireball. Lightning can also create electromagnetic fields that can interfere with electrical equipment, such as radios, televisions, and computers.

In summary, thunder is the sound that results from a lightning strike, which ionizes the air and creates a channel of plasma through which the electric current flows. The intense heat of the lightning bolt causes the air to rapidly expand and contract, creating a shock wave that travels through the air and creates the sound of thunder. While thunder can be a fascinating natural phenomenon to observe, it is also important to remember that lightning can be dangerous and should be taken seriously.

17

When should we purchase gasoline for our car during the day?

There is a popular belief that buying gasoline for your car during certain times of the day can save you money. According to this belief, gasoline is denser and thus contains more energy during cooler parts of the day, such as early morning or late evening. However, is there any truth to this belief? Let's explore the science behind it.

Firstly, gasoline is made up of a complex mixture of hydrocarbons, which are compounds containing hydrogen and carbon. The density of gasoline varies slightly depending on its temperature, with colder gasoline being slightly more dense than warmer gasoline. However, the difference in density between gasoline at different temperatures is relatively small and typically only affects the volume of gasoline you get per gallon, rather than the energy content.

Secondly, the energy content of gasoline is measured in British Thermal Units (BTUs). The amount of energy released when a gallon of gasoline is burned remains constant regardless of the temperature of the gasoline. This means that whether you buy gasoline in the early morning or late evening, the amount of energy you get per gallon will be the same.

Furthermore, gasoline is stored underground in tanks at gas stations, which helps to regulate its temperature. Gasoline pumps also have filters that prevent sediment and water from getting into your gas tank. These filters may become clogged if gasoline is drawn from the bottom of the storage tank where sediment and water may accumulate. However, this is not related to the time of day at which gasoline is pumped.

In conclusion, there is no significant difference in the energy content of gasoline purchased at different times of the day. While cooler gasoline may be slightly more dense, the difference in energy content is negligible. Therefore, there is no need to worry about buying gasoline at a specific time of day to save money or improve your car's performance. It is more important to focus on factors such as finding a gas station with competitive prices and regularly maintaining your car's engine to ensure optimal fuel efficiency.

18

We say that the reactivity of metals increases from top to bottom on the periodic table, but what is the reactivity sequence of metals, such as Li, K, Ba, Sr...?

The reactivity of metals refers to their tendency to form chemical bonds with other elements, particularly oxygen. Generally, metals at the top of the periodic table are more reactive than those at the bottom. This is due to the fact that the outermost electrons of these metals are more loosely bound to the nucleus, making them easier to remove and allowing them to react more readily with other elements.

However, the reactivity of specific metals can vary depending on their chemical properties and the conditions under which they are reacting. In the case of the metals Li, K, Ba, and Sr, their reactivity follows a specific sequence.

Lithium (Li) is the most reactive of the four metals, followed by potassium (K), barium (Ba), and strontium (Sr). This is due to the fact that lithium has the smallest atomic radius and the strongest electronegativity, meaning its outermost electrons are held more tightly to the nucleus and are more difficult to remove. As a result, lithium reacts more readily with other elements.

Potassium is the second most reactive metal in this sequence. Like lithium, potassium has a relatively small atomic radius, which makes its outermost electrons more tightly bound to the nucleus. However, potassium has a lower electronegativity than lithium, which means that its outermost electrons are slightly more loosely bound and it is more reactive than some other metals.

Barium and strontium are less reactive than lithium and potassium, but still more reactive than many other metals. This is due to their relatively small atomic radii, which make their outermost electrons somewhat more tightly bound to the nucleus. However, they have a lower electronegativity than lithium and potassium, which means that their outermost electrons are slightly more loosely bound and they are more reactive than some other metals.

It is important to note that the reactivity of metals can also depend on the specific conditions under which they are reacting. For example, in the presence of water or air, some metals can form a protective oxide layer on their surface, which can reduce their reactivity. Additionally, the presence of other chemicals or impurities in the reaction environment can also impact the reactivity of metals.

In summary, the reactivity sequence of the metals Li, K, Ba, and Sr follows the order of lithium > potassium > barium > strontium. This is due to a combination of factors, including their atomic radius, electronegativity, and the specific conditions under which they are reacting. Understanding the reactivity sequence of metals is important for predicting and controlling chemical reactions and for selecting appropriate materials for various applications.

19

What is the pH and What is the pH of Our Body ?

The pH is a measure of the acidity or basicity of a solution, expressed on a scale of 0 to 14. A solution with a pH of 7 is considered neutral, while solutions with pH values lower than 7 are considered acidic and those with pH values higher than 7 are considered basic.

The pH of our body is tightly regulated, as even small deviations from the normal range can have serious health consequences. The pH of human blood, for example, is maintained within a narrow range of 7.35 to 7.45, which is slightly alkaline. This is achieved through the action of various buffering systems in the body, which help to resist changes in pH.

The pH of different parts of the body can vary, however. For example, the pH of the stomach is highly acidic, with a pH range of 1.5 to 3.5, which is necessary for the proper digestion of food. The pH of saliva, on the other hand, is slightly acidic, with a pH range of 6.5 to 7.5, which helps to protect the teeth from decay.

The pH of the skin also varies depending on the location on the body. The skin on the face, for example, has a pH of around 5.5, which is slightly acidic, while the skin on the scalp can have a pH of up to 7.5, which is slightly basic.

The importance of pH regulation in the body cannot be overstated. Many biochemical reactions, such as enzyme activity, are highly sensitive to changes in pH, and even small deviations from the normal range can disrupt these reactions and lead to health problems.

In addition to the pH of our body, the pH of various substances we encounter in our daily lives can also have an impact on our health. For example, acidic beverages such as soda can erode tooth enamel over time, while exposure to basic substances such as bleach can cause chemical burns.

Overall, the pH is an important concept in understanding the chemical properties of substances and their effects on the body. By maintaining a proper pH balance, our bodies can function properly and avoid many potential health problems.

20

What does it mean if say one mole of hydrogen molecules?

A mole is a unit of measurement used in chemistry to represent a large number of particles, such as atoms, molecules, or ions. One mole of a substance is equal to Avogadro's number, which is approximately 6.022×10^{23} particles. Therefore, if we say one mole of hydrogen molecules, we mean 6.022×10^{23} molecules of hydrogen.

Hydrogen is the simplest and most abundant element in the universe, with its atomic number being 1. It exists in nature mainly in the form of diatomic molecules, H_2, which consist of two hydrogen atoms bonded together by a covalent bond. One mole of hydrogen molecules, therefore, contains 6.022×10^{23} H_2 molecules.

The concept of the mole is useful in chemical calculations and measurements because it allows chemists to measure the quantity of a substance in a more meaningful way than simply counting individual particles. For example, when calculating the amount of reactants needed to produce a specific amount of product in a chemical reaction, chemists use the mole concept to ensure that the reactants are added in the correct proportions.

It is also important to note that the mole concept is closely related to the atomic mass unit (amu) of elements. The atomic mass unit is defined as 1/12th the mass of one carbon-12 atom, which has 6 protons and 6 neutrons in its nucleus. The atomic mass unit is used to measure the mass of individual atoms and molecules, and the mass of one mole of a substance in grams is equal to its molar mass. For example, the molar mass of hydrogen gas is approximately 2 grams per mole, which means that one mole of hydrogen gas weighs 2 grams.

Regarding the application of the mole concept, it is used in many branches of chemistry, including analytical chemistry, physical chemistry, and organic chemistry, to name a few. It is also a fundamental concept in stoichiometry, which is the study of the quantitative relationships between reactants and products in a chemical reaction.

In summary, when we say one mole of hydrogen molecules, we are referring to 6.022×10^{23} H2 molecules. The mole concept is essential in chemical calculations and allows chemists to measure the quantity of a substance in a meaningful way. The concept is widely used in many branches of chemistry and is fundamental in stoichiometry.

21

Which compound, known asaqua fortis and spirit of niter, is used in everyday life as an acid?

The compound known as aqua fortis or spirit of niter is nitric acid (HNO_3), a highly corrosive and toxic acid used in various industries and applications. Nitric acid is one of the most common and important industrial chemicals, with a wide range of uses including the production of fertilizers, explosives, dyes, and pharmaceuticals.

Nitric acid is a strong acid, meaning that it dissociates almost completely in water to produce hydrogen ions ($H+$) and nitrate ions (NO_3-). This high concentration of hydrogen ions gives nitric acid its acidic properties, such as its ability to react with metals and neutralize bases.

In everyday life, nitric acid is commonly used as a laboratory reagent for various analytical and synthetic procedures. It is also used in the production of stainless steel, semiconductors, and rocket propellants. Additionally, nitric acid is used in the cleaning and etching of metals and glass, as well as in the manufacture of explosives, such as dynamite and TNT.

Despite its many uses, nitric acid is also a hazardous substance that can cause serious health effects and environmental damage if not handled properly. The concentrated acid is corrosive to skin, eyes, and respiratory tract, and can cause severe burns upon contact. Inhalation of nitric acid fumes can cause irritation to the lungs and throat, and long-term exposure may lead to respiratory problems and cancer.

In conclusion, nitric acid is a common and important acid used in various industrial and laboratory applications, as well as in everyday life. However, its corrosive and toxic nature requires careful handling and proper disposal to ensure the safety of workers and the environment.

22

What are the chemicals used in the production of batteries and accumulators?

Batteries and accumulators are widely used in various electronic devices and transportation systems. The chemical composition of batteries and accumulators can vary depending on their type and intended use, but they all involve the use of various chemicals in their production. In this article, we will explore the different chemicals used in the production of batteries and accumulators.

1. Lithium-ion batteries: Lithium-ion batteries are widely used in portable electronic devices, electric vehicles, and renewable energy systems. These batteries contain a cathode (positive electrode), an anode (negative electrode), and an electrolyte that separates the electrodes. The cathode is typically made of lithium cobalt oxide ($LiCoO_2$), lithium manganese oxide ($LiMn_2O_4$), or lithium iron phosphate ($LiFePO_4$), while the anode is made of graphite. The electrolyte is a lithium salt in an organic solvent such as ethylene carbonate or dimethyl carbonate.
2. Lead-acid batteries: Lead-acid batteries are commonly used in automotive, marine, and uninterruptible power supply (UPS) applications. These batteries consist of lead plates as electrodes and sulfuric acid as the electrolyte. The lead plates are coated with lead dioxide (PbO_2) as the

positive electrode and lead (Pb) as the negative electrode. The sulfuric acid acts as a conductor of ions between the two electrodes.

3. Nickel–cadmium batteries: Nickel–cadmium batteries are widely used in portable electronic devices, power tools, and emergency lighting systems. These batteries contain a nickel oxide hydroxide (NiOOH) cathode and a cadmium (Cd) anode. The electrolyte is a potassium hydroxide (KOH) solution.

4. Nickel-metal hydride batteries: Nickel-metal hydride batteries are commonly used in hybrid electric vehicles, portable electronic devices, and emergency lighting systems. These batteries contain a nickel oxide hydroxide (NiOOH) cathode, a hydrogen–absorbing alloy as the anode, and a potassium hydroxide (KOH) electrolyte.

5. Zinc-carbon batteries: Zinc-carbon batteries are commonly used in small electronic devices such as flashlights, toys, and remote controls. These batteries consist of a zinc (Zn) anode, a manganese dioxide (MnO_2) cathode, and an electrolyte made of ammonium chloride (NH_4Cl) and zinc chloride ($ZnCl_2$).

6. Lithium polymer batteries: Lithium polymer batteries are commonly used in portable electronic devices and drones. These batteries use a polymer electrolyte instead of a liquid electrolyte, which makes them lighter and more flexible than lithium–ion batteries. The cathode and anode materials are similar to those used in lithium–ion batteries.

In conclusion, batteries and accumulators are essential components of modern technology, and their production involves the use of various chemicals such as lithium, lead, nickel, cadmium, zinc, and manganese. The choice of chemicals depends on the intended use and performance requirements of the battery or accumulator. As the demand for renewable energy and electric vehicles continues to grow, the production of batteries and accumulators will become increasingly important to meet the needs of society.

23

How does soap clean?

Soap is a common household cleaning agent that is used to remove dirt, oil, and other impurities from a variety of surfaces. But how does soap work to clean these surfaces? In this article, we will explore the science behind soap and its cleaning power.

To understand how soap works, we first need to understand the nature of the substances that need to be cleaned. Dirt, oil, and other impurities are generally hydrophobic, meaning they are repelled by water. This makes it difficult to remove them using water alone. Soap, on the other hand, is made up of molecules that have both hydrophilic (water-loving) and hydrophobic (water-repelling) parts. These molecules are called amphiphiles.

When soap is mixed with water, the hydrophobic tails of the soap molecules are drawn to the dirt and oil on the surface being cleaned. This breaks up the clusters of dirt and oil, allowing them to be surrounded by the hydrophilic heads of the soap molecules. These heads then form a shell around the dirt and oil, called a micelle, which can be easily rinsed away with water.

Soap is also effective at removing germs and bacteria from surfaces. The hydrophobic tails of the soap molecules can penetrate the fatty membranes of bacteria and viruses, causing them to break apart and be washed away.

It is important to note that not all types of soap are created equal when it comes to cleaning power. The effectiveness of soap depends on factors such as the type and amount of surfactants (the active cleaning agents) used, as well as the water temperature and pH level. For example, some soaps are designed to be more effective at removing grease, while others are better at removing stains.

In addition to its cleaning power, soap also has some other useful properties. It can help to soften hard water by binding to the minerals that cause it to be hard, which can improve its cleaning ability. Soap also has a low toxicity and is generally safe for use on most surfaces and materials.

In conclusion, soap is an effective cleaning agent that works by using amphiphilic molecules to break up clusters of dirt and oil, and form micelles that can be easily rinsed away with water. The effectiveness of soap depends on several factors, including the type and amount of surfactants used, as well as water temperature and pH level. With its many useful properties, soap is a staple cleaning agent that is used in households and industries around the world.

24

Why is seawater salty?

Seawater is salty because it contains various dissolved salts and minerals. The ocean is the largest body of water on the planet, and it contains an estimated 50 quadrillion tons of salt. But why is seawater salty? The answer to this question is complex and multifaceted, as there are several different factors that contribute to the saltiness of the ocean.

One of the main reasons why seawater is salty is because of the Earth's geology. Over millions of years, minerals from the Earth's crust have eroded and washed into the ocean, where they have dissolved into the water. These minerals include compounds such as sodium chloride (table salt), magnesium, calcium, and potassium, all of which contribute to the salinity of seawater.

Another factor that contributes to the saltiness of seawater is the water cycle. When water evaporates from the ocean's surface, it leaves behind the salt and minerals that were dissolved in it. This process is called "evaporative concentration," and it helps to concentrate the salt in the ocean. As the concentrated water vapor rises into the atmosphere and cools, it forms clouds, which eventually release their moisture as rain or snow. When this precipitation falls on land, it can dissolve minerals from rocks and soil, which then flow into rivers and eventually make their way back into the ocean.

The ocean's currents also play a role in the saltiness of seawater. As seawater circulates around the world, it carries with it the salt and minerals that were dissolved in it. In some areas of the ocean, such as the Atlantic Ocean, there are regions where the surface water becomes so salty that it sinks to the bottom, creating what is known as a "halocline." These areas of concentrated salt can have a significant impact on the overall salinity of the ocean.

Finally, the amount of rainfall and freshwater that enters the ocean also plays a role in the saltiness of seawater. In areas where there is high rainfall or freshwater runoff, such as near river deltas, the salinity of the ocean can be diluted. Conversely, in areas where there is little rainfall and a high rate of evaporation, such as the Red Sea, the salinity can be much higher than average.

In summary, seawater is salty because of a combination of factors, including the Earth's geology, the water cycle, ocean currents, and the amount of freshwater that enters the ocean. While the saltiness of seawater can vary depending on location, these factors ensure that the ocean remains a highly concentrated source of minerals and salts.

25

Which element has the highest malleability?

The element with the highest malleability is gold (Au).

Malleability is the ability of a material to deform under compression, bending or stretching forces without cracking or breaking. In other words, it is the property of a substance to be easily shaped into thin sheets by hammering or rolling without losing its strength. Malleability is an important characteristic of metals and is determined by the arrangement of atoms in a crystal lattice.

Gold is the most malleable metal, which means it can be easily hammered or rolled into thin sheets without breaking or cracking. It is so malleable that one gram of gold can be beaten into a sheet of 1 square meter. This property is due to its unique electronic structure, which makes it easy for the atoms to move and adjust their positions when subjected to an external force. In fact, gold is so malleable that it can be beaten into a sheet that is just a few atoms thick.

Other metals that are highly malleable include silver (Ag), copper (Cu), and aluminum (Al). However, gold has the highest malleability of all metals, making it the preferred choice for many applications that require thin, flexible sheets, such as gold leaf, jewelry, and electrical contacts.

The malleability of gold has been recognized and valued by humans for thousands of years. Gold artifacts have been found dating back to ancient civilizations, such as the Egyptians, who used gold to make decorative objects and jewelry. The use of gold as a currency is also well documented throughout history. Today, gold remains a highly valued commodity, both for its malleability and its other unique properties, such as its luster, rarity, and resistance to corrosion.

26

How does the process of adhesion occur?

Adhesion is a process in which two different surfaces are attracted and stick together. This process can occur between similar or dissimilar substances, such as solids, liquids, or gases. Adhesion is a critical process in many natural and artificial systems, such as cell adhesion, adhesive bonding, and surface coating.

The adhesion process occurs due to the interaction between the molecules on the surfaces of the two materials. The molecules on the surfaces are attracted to each other through various intermolecular forces, such as Van der Waals forces, hydrogen bonding, dipole-dipole interactions, and electrostatic forces.

Van der Waals forces are the weakest intermolecular forces that exist between all molecules. These forces are caused by fluctuations in the electron distribution of the molecules, which cause temporary dipoles that can attract other molecules. Hydrogen bonding occurs when a hydrogen atom in a molecule is bonded to a highly electronegative atom, such as oxygen or nitrogen. The hydrogen atom in one molecule is attracted to the electronegative atom in another molecule, forming a hydrogen bond. Dipole-dipole interactions occur between polar molecules, which have a permanent dipole moment due to an asymmetric distribution of electrons. Electrostatic forces are the strongest intermolecular forces and occur between charged molecules.

The adhesion process can be affected by several factors, such as the nature of the two surfaces, the roughness of the surfaces, the temperature, the humidity, and the duration of contact. The nature of the surfaces can affect the strength of the adhesion, as different materials have different intermolecular forces. For example, a polar surface will adhere better to another polar surface due to the presence of dipole-dipole and hydrogen bonding interactions. The roughness of the surfaces can also affect adhesion, as it can increase the contact area between the two surfaces, allowing more intermolecular forces to come into play.

Temperature and humidity can also affect adhesion, as they can change the properties of the surfaces, such as their polarity and roughness. A higher temperature can increase the mobility of the molecules on the surfaces, allowing them to interact more easily with each other. A higher humidity can increase the surface energy of the surfaces, making them more attractive to each other.

The duration of contact is also important for adhesion, as it can affect the strength of the intermolecular forces between the two surfaces. A longer contact time can allow more intermolecular forces to develop, resulting in stronger adhesion.

In conclusion, adhesion is a complex process that occurs due to the interaction between the molecules on the surfaces of two materials. This process is critical in many natural and artificial systems, and its strength can be affected by several factors, such as the nature of the surfaces, the roughness of the surfaces, the temperature, the humidity, and the duration of contact. Understanding the adhesion process can help in the development of new materials and technologies.

27

Can you provide information about the amounts of some elements found in the human body?

The human body is composed of a complex mixture of elements, but some are more abundant than others. Here are the approximate amounts of some of the most common elements found in the human body:

1. Oxygen (O) - 65% Oxygen is the most abundant element in the human body, making up almost two-thirds of the body's mass. It is a crucial component of water and many other organic molecules that make up living organisms.

2. Carbon (C) - 18% Carbon is the second most abundant element in the human body and is found in all organic molecules, including proteins, fats, and carbohydrates. It is the backbone of life and is essential for the formation of complex molecules.

3. Hydrogen (H) - 10% Hydrogen is the third most abundant element in the human body and is a component of water, which makes up a significant portion of the body's mass. It is also found in all organic molecules, along with carbon.

4. Nitrogen (N) - 3% Nitrogen is a component of proteins and nucleic

acids, which are the building blocks of life. It is also found in many other organic molecules, including amino acids and nucleotides.

5. Calcium (Ca) – 1.5% Calcium is an important mineral that is essential for bone and teeth health. It also plays a role in muscle function and nerve transmission.

6. Phosphorus (P) – 1% Phosphorus is a component of nucleic acids, which make up the genetic material of cells. It is also important for bone and teeth health and is involved in energy metabolism.

7. Potassium (K) – 0.25% Potassium is an essential mineral that is involved in many physiological processes, including muscle function and nerve transmission.

8. Sulfur (S) – 0.25% Sulfur is a component of many proteins and is involved in the formation of disulfide bonds, which help to stabilize protein structures.

9. Sodium (Na) – 0.15% Sodium is an essential mineral that is involved in many physiological processes, including fluid balance and nerve transmission.

10. Chlorine (Cl) – 0.15% Chlorine is a component of salt, which is important for fluid balance and is involved in many physiological processes.

These are just a few of the most abundant elements found in the human body. Other elements, such as iron, zinc, and copper, are present in smaller amounts but are also essential for human health.

28

What is a mirage, and why do small puddles appear at a certain distance on a hot road on a sunny day and disappear before reaching them?

A mirage is a visual phenomenon that occurs due to the bending of light rays in the atmosphere. It causes an optical illusion, making distant objects appear distorted or displaced. Mirages are often seen on hot, sunny days, especially over hot surfaces such as deserts, where the temperature gradient in the atmosphere is high.

The appearance of small puddles on a hot road on a sunny day is an example of a mirage. This phenomenon is called a "mirage puddle" or "Fata Morgana." It occurs because of the refraction of light rays passing through the different layers of air with different temperatures and densities above the surface of the road. The hot surface of the road heats the air above it, creating a layer of hot air close to the surface. This hot layer of air has a lower density than the cooler air above it, which has a higher density.

When light rays from the sky pass through these layers of air with different densities and temperatures, they bend, refract, and reflect in different directions. The bending of light rays causes an optical illusion, making the surface of the road appear to be wet, with small puddles of water. However, these puddles are not real; they are a result of the refraction of light.

The Fata Morgana mirage is named after Morgan le Fay, a sorceress in Arthurian legend who was said to have created illusions. It is a complex mirage that can produce images of objects that are both upside down and right side up. The Fata Morgana mirage occurs when there is a temperature inversion in the atmosphere, where the temperature increases with height instead of decreasing.

In addition to mirage puddles and Fata Morgana, other types of mirages include superior and inferior mirages. A superior mirage occurs when a distant object appears to be floating above its actual position, while an inferior mirage makes an object appear lower than its actual position.

In conclusion, a mirage is a visual phenomenon caused by the bending of light rays in the atmosphere. The appearance of small puddles on a hot road on a sunny day is an example of a mirage puddle or Fata Morgana, which occurs due to the refraction of light rays passing through layers of air with different densities and temperatures. These puddles are not real and are just an optical illusion.

29

What is an atomic bomb?

An atomic bomb, also known as a nuclear bomb, is a weapon of mass destruction that uses nuclear reactions to release an enormous amount of energy in the form of an explosion. The explosion of an atomic bomb is caused by the rapid and uncontrolled release of energy from the nucleus of an atom. Atomic bombs are capable of causing massive destruction and devastation, with the potential to kill millions of people and destroy entire cities.

The first atomic bomb was developed by the United States during World War II as part of the Manhattan Project. The bomb was dropped on the Japanese cities of Hiroshima and Nagasaki in August 1945, leading to the deaths of over 200,000 people, mostly civilians. These bombings remain the only use of atomic weapons in warfare to this day.

The basic principle behind an atomic bomb is nuclear fission, which involves splitting the nucleus of an atom into two smaller nuclei, releasing a large amount of energy in the process. This energy is released in the form of gamma rays, which are high-energy photons that can cause damage to living organisms.

The fuel used in an atomic bomb is usually uranium-235 or plutonium-239, which are both radioactive and can undergo nuclear fission. When a neutron collides with a uranium-235 or plutonium-239 nucleus, it causes the nucleus to become unstable and split into two smaller nuclei, releasing a large amount of energy and more neutrons. These neutrons can then collide with other nuclei, causing a chain reaction that leads to the rapid release of energy and a nuclear explosion.

The explosion of an atomic bomb releases an enormous amount of energy in the form of heat, light, and radiation. The heat from the explosion can cause fires and burn everything in its path, while the blast wave can cause severe damage to structures and infrastructure. The radiation released by the explosion can cause radiation sickness and increase the risk of cancer and other diseases.

Due to the destructive power of atomic bombs, there have been efforts to limit their spread and use. The United Nations adopted the Treaty on the Non-Proliferation of Nuclear Weapons in 1968, which aims to prevent the spread of nuclear weapons and promote peaceful use of nuclear energy. Many countries have signed the treaty, but some, including North Korea, have developed nuclear weapons despite international efforts to prevent their acquisition.

In summary, an atomic bomb is a weapon of mass destruction that uses nuclear reactions to release an enormous amount of energy in the form of an explosion. The explosion is caused by the rapid and uncontrolled release of energy from the nucleus of an atom, and the fuel used is usually uranium-235 or plutonium-239. The destructive power of atomic bombs has led to efforts to limit their spread and use, but they remain a significant threat to global security.

30

Antifreeze fish

Antarctic fish are known for their ability to survive in the frigid waters of the Southern Ocean, where temperatures can drop below freezing. These fish have developed several unique adaptations to help them survive in these extreme conditions, one of which is the presence of antifreeze proteins in their blood.

Antifreeze proteins, or AFGPs, are a type of glycoprotein that bind to ice crystals and prevent them from growing larger. AFGPs work by adsorbing to the surface of ice crystals, which prevents water molecules from joining the crystal lattice and ultimately stops ice crystal growth. This allows the fish to survive in subzero temperatures by preventing the formation of ice crystals in their tissues.

There are several types of antifreeze proteins, but the AFGPs found in Antarctic fish are particularly effective at preventing ice crystal growth. These proteins are produced in the liver and secreted into the bloodstream, where they can be transported to all parts of the body. The concentration of AFGPs in the blood varies depending on the species of fish and the conditions in which they live.

The presence of antifreeze proteins in Antarctic fish was first discovered in the late 1960s by Arthur DeVries, a biochemist at the University of Illinois. DeVries was studying the physiology of Antarctic fish and noticed that they had high levels of a substance in their blood that prevented ice formation. He isolated the protein responsible and named it "antifreeze glycoprotein."

Since then, researchers have been studying the properties and structure of AFGPs to better understand how they work. AFGPs are made up of a repetitive sequence of amino acids, with several sugar molecules attached. The exact sequence of amino acids varies between different species of fish, but all AFGPs have a similar structure and function.

AFGPs are not only found in fish, but also in other organisms that live in cold environments, such as insects, plants, and bacteria. Researchers are studying the potential applications of AFGPs in a variety of fields, including medicine and food preservation.

In conclusion, the presence of antifreeze proteins in Antarctic fish is a remarkable adaptation that allows these fish to survive in extremely cold temperatures. AFGPs prevent the formation of ice crystals in their tissues, which would otherwise be fatal. The study of these proteins has led to a better understanding of how organisms adapt to extreme environments and has potential applications in a variety of fields.

31

Some acids and salts used in daily life

Acids and salts are commonly used in our daily lives for various purposes. From food preservation to cleaning and medical treatments, they play a significant role. In this article, we will discuss some of the commonly used acids and salts and their applications.

1. Acetic Acid Acetic acid, also known as vinegar, is a weak acid commonly used in food preservation and cooking. It is used as a preservative for pickles, as a condiment for salads and as a flavor enhancer. It is also used in the production of dyes, plastics, and textiles.
2. Citric Acid Citric acid is a weak organic acid found in citrus fruits such as lemons and oranges. It is used as a food additive, flavor enhancer, and preservative. It is also used in the production of detergents, cosmetics, and pharmaceuticals.
3. Sulfuric Acid Sulfuric acid is a strong acid commonly used in the production of fertilizers, dyes, detergents, and explosives. It is also used in the refining of metals and as a laboratory reagent.
4. Hydrochloric Acid Hydrochloric acid is a strong acid used in the production of various chemicals such as PVC, polyurethane, and rubber. It is also used in the pickling of steel, as a laboratory reagent, and as a cleaning agent.
5. Sodium Chloride Sodium chloride, also known as table salt, is a common

salt used for seasoning and food preservation. It is also used in the production of various chemicals such as chlorine, sodium hydroxide, and sodium carbonate.

6. Calcium Carbonate Calcium carbonate is a salt commonly used in the production of cement, glass, and ceramics. It is also used as a dietary supplement and antacid.

7. Sodium Bicarbonate Sodium bicarbonate, also known as baking soda, is a salt commonly used in baking as a leavening agent. It is also used as a cleaning agent and in medical treatments such as antacid and as a urinary alkalinizer.

8. Potassium Nitrate Potassium nitrate, also known as saltpeter, is a salt commonly used in the production of fertilizers, gunpowder, and fireworks. It is also used in the preservation of meat and as a component in toothpaste for sensitive teeth.

9. Sodium Fluoride Sodium fluoride is a salt commonly used in toothpaste and mouthwash for dental health. It is also used in water fluoridation programs to prevent tooth decay.

10. Ammonium Nitrate Ammonium nitrate is a salt commonly used as a fertilizer and in the production of explosives such as dynamite. It is also used in the mining industry for blasting and in the production of nitrogen-rich compounds.

In conclusion, acids and salts have numerous applications in our daily lives. From food preservation to cleaning and medical treatments, they play an important role in various industries. It is important to use these substances carefully and responsibly to avoid any potential harm or damage.

32

Airbag

Airbags are a crucial safety feature in cars that can prevent serious injury or even death in the event of an accident. The chemistry behind airbags is fascinating, and in this article, we will explore the science behind how they work.

Airbags were first introduced in the 1970s, and since then, they have become an essential part of car safety. They are designed to deploy in the event of a crash, providing a cushioning effect to protect the driver and passengers from the force of impact. The chemical reactions that take place inside an airbag are what make this possible.

Airbags contain a mixture of chemicals, including sodium azide, potassium nitrate, and silica. Sodium azide is a highly reactive compound that is used as the primary propellant in airbags. When the airbag is triggered, an electrical charge is sent to the initiator, which ignites the sodium azide, causing it to rapidly decompose into nitrogen gas and sodium metal.

The reaction is highly exothermic, meaning it produces a large amount of heat and gas in a short amount of time. The nitrogen gas quickly expands, filling the airbag and causing it to inflate. The sodium metal produced by the reaction is contained within the airbag, preventing it from reacting with other chemicals or causing damage to the car or passengers.

In addition to sodium azide, airbags also contain potassium nitrate and silica. Potassium nitrate is added to the mixture to help regulate the rate of the reaction. Without it, the reaction would be too fast and violent, potentially causing injury to the passengers. Silica is used as a flow agent, helping to distribute the chemicals evenly within the airbag.

One of the main challenges in designing airbags is ensuring that they deploy at the right time and with the right amount of force. Too much force could cause injury to the passengers, while too little force could fail to protect them in the event of a crash. The amount of force required depends on various factors, such as the speed of the car and the angle of impact.

Modern airbags are equipped with sensors that detect when a crash has occurred and determine the appropriate amount of force needed to inflate the airbag. This is done using sophisticated algorithms that take into account various factors such as the weight and position of the passengers.

In conclusion, airbags are a crucial safety feature in cars that rely on the chemistry of sodium azide and other compounds to rapidly inflate and cushion passengers in the event of an accident. The careful design of airbags involves balancing the need for rapid deployment with the need to prevent injury to passengers, making use of advanced sensors and algorithms to ensure that they deploy at the right time and with the right amount of force.

33

The chemistry of what we eat

The food we eat is composed of a variety of chemicals, and understanding the chemistry behind what we consume can help us make informed choices about our diets. From carbohydrates and proteins to vitamins and minerals, each component of our food has its own unique chemistry that plays a role in our overall health and wellbeing.

Carbohydrates are one of the primary components of our diet, and they play a crucial role in providing our bodies with energy. Carbohydrates are composed of molecules made up of carbon, hydrogen, and oxygen atoms, and they come in two main forms: simple and complex. Simple carbohydrates, such as sugar, are made up of just one or two sugar molecules and are quickly broken down by the body, while complex carbohydrates, such as whole grains, are made up of long chains of sugar molecules that take longer to digest.

Proteins are another important component of our diet, and they are composed of amino acids. There are 20 different amino acids, and they can be combined in different ways to create a wide variety of proteins with unique functions. Proteins are essential for building and repairing tissues in the body, and they also play a role in the production of enzymes and hormones.

Fats are another important component of our diet, and they play a role in providing our bodies with energy as well as helping to absorb and transport vitamins. Fats are composed of molecules called fatty acids, which can be either saturated or unsaturated. Saturated fats, which are found in animal products such as butter and cheese, are solid at room temperature and are linked to an increased risk of heart disease. Unsaturated fats, which are found in foods such as nuts and olive oil, are liquid at room temperature and can help to lower cholesterol levels in the body.

Vitamins and minerals are also important components of our diet, and they play a role in a wide variety of bodily functions. Vitamins are organic compounds that the body needs in small amounts, while minerals are inorganic compounds that are required for a range of bodily functions, such as building strong bones and teeth. Some of the most important vitamins and minerals include vitamin C, vitamin D, calcium, and iron.

In addition to these basic components of our diet, there are also a variety of additives and preservatives that are commonly found in the foods we eat. These include things like artificial sweeteners, food dyes, and sodium benzoate. While many of these compounds are considered safe in small amounts, some people may have sensitivities or allergies to certain additives, and it is important to read food labels carefully and make informed choices about what we eat.

Overall, the chemistry of what we eat is complex and multifaceted, and there is still much to be learned about how different components of our diets interact with our bodies. By understanding the basic chemistry behind the foods we consume, however, we can make informed choices about our diets and take steps to ensure that we are giving our bodies the nutrients they need to function at their best.

34

History of chemistry

Chemistry is the branch of science that deals with the study of matter, its properties, structure, composition, and behavior. It has a long and fascinating history that dates back to ancient times, and its evolution over the centuries has led to numerous scientific discoveries and advancements.

The history of chemistry can be traced back to ancient civilizations, such as Egypt, China, and India, where alchemy was practiced as a form of natural philosophy. Alchemists were interested in transmuting base metals into gold, discovering the philosopher's stone, and creating elixirs of immortality.

During the Middle Ages, the Islamic world played a significant role in the development of chemistry. Scholars like Jabir ibn Hayyan (also known as Geber) introduced experimental methods and improved the techniques of distillation and sublimation. They also discovered many chemical substances, including alcohol, sulfuric acid, and nitric acid.

In Europe, alchemy continued to be practiced until the 17th century. However, the emergence of modern chemistry as a scientific discipline began with the work of the famous English chemist Robert Boyle. In his book "The Sceptical Chymist" published in 1661, Boyle challenged the Aristotelian notion of the four elements and introduced the idea of chemical elements.

Antoine Lavoisier, a French chemist, is often referred to as the father of modern chemistry. He established the law of conservation of mass and the law of definite proportions and introduced the concept of the chemical reaction. He also discovered oxygen and named it, and demonstrated its role in combustion and respiration.

In the 19th century, chemistry expanded rapidly with the discovery of new elements, such as helium, neon, and argon, and the development of the periodic table by Dmitri Mendeleev. Organic chemistry, the study of carbon-based compounds, also emerged as a separate branch of chemistry during this time.

The 20th century saw significant advancements in the field of chemistry, including the development of quantum mechanics and its application to chemical systems. This led to a better understanding of chemical bonding, molecular orbitals, and electronic structure. Synthetic chemistry also flourished, leading to the discovery of numerous new compounds and materials.

Today, chemistry plays a crucial role in many aspects of modern life, from the production of medicines and materials to the development of alternative energy sources and environmental protection. The field is constantly evolving, with new discoveries and innovations being made every day.

In conclusion, the history of chemistry is long and rich, spanning thousands of years and numerous civilizations. From the ancient alchemists to modern-day scientists, chemists have made significant contributions to our understanding of the natural world and the materials we use every day. The study of chemistry continues to evolve and shape our world in profound ways.

35

How long do garbage dumps last?

Garbage dumps, also known as landfills, are designed to hold and contain solid waste materials generated by human activities. The amount of time a landfill lasts varies depending on several factors, including the type and amount of waste, the landfill design and management practices, and the local climate conditions.

Typically, the decomposition of organic waste is the primary factor determining the longevity of a landfill. Organic waste, such as food scraps, yard waste, and paper products, can take years to decompose in a landfill. The decomposition process can be slowed down by the lack of oxygen, moisture, and sunlight in the landfill environment. This results in the accumulation of large amounts of waste that can take up significant space in the landfill.

According to the Environmental Protection Agency (EPA), the average lifespan of a landfill in the United States is approximately 30 years. However, this can vary greatly depending on the factors mentioned above. In some cases, a landfill may continue to receive waste long after it is officially closed, which can extend its lifespan.

One of the biggest concerns with landfills is the production of methane gas, a potent greenhouse gas that contributes to climate change. As organic waste decomposes, it produces methane gas that can escape into the atmosphere. Landfills are required to have systems in place to capture and control methane emissions, which can be used as a source of renewable energy.

To mitigate the environmental impact of landfills, many countries are exploring alternative waste management practices such as recycling, composting, and waste-to-energy facilities. These practices can divert waste from landfills and reduce the amount of waste that needs to be disposed of.

In conclusion, the lifespan of a landfill varies based on a variety of factors and can range from several years to several decades. While landfills are an essential part of our waste management infrastructure, they also present significant environmental challenges that must be addressed. It is important to continue exploring alternative waste management practices to reduce our reliance on landfills and minimize their impact on the environment.

36

How did Napoleon die?

Napoleon Bonaparte, one of the greatest military commanders in history, died on May 5, 1821, at the age of 51. The cause of his death has been the subject of much speculation and debate over the years.

Napoleon was exiled to the remote island of Saint Helena in the South Atlantic after his defeat at the Battle of Waterloo in 1815. He spent the last six years of his life there, confined to a small house with a few loyal followers.

There are several theories about how Napoleon died, but the most widely accepted one is that he died of stomach cancer. According to his doctor, Francesco Antommarchi, Napoleon began suffering from stomach pain in 1820. His condition worsened over time, and he began experiencing other symptoms such as vomiting, diarrhea, and weight loss. Antommarchi diagnosed him with cancer of the stomach and liver.

Some historians have suggested that Napoleon may have been poisoned with arsenic, either by his captors or by someone in his entourage. This theory is based on the fact that high levels of arsenic were found in his hair after his death. However, modern medical experts have cast doubt on this theory, arguing that the levels of arsenic found in his hair were not high enough to have caused his death.

Another theory is that Napoleon died of a heart attack. This theory is based on the fact that he had a history of heart problems and that his death was sudden and unexpected. However, there is little evidence to support this theory, and most medical experts believe that he died of cancer.

In any case, Napoleon's death was a great loss for France and for the world. He was one of the most brilliant and innovative military leaders in history, and his legacy can still be felt today in many areas of modern life.

37

What is the anesthetic gas used in surgery?

The use of anesthesia has revolutionized surgery by enabling complex procedures to be performed safely and effectively. Anesthesia refers to a state of temporary unconsciousness or insensitivity to pain, which is induced by administering one or more anesthetic drugs. One common type of anesthesia used in surgery is general anesthesia, which is induced by administering anesthetic gases.

The most commonly used anesthetic gas in surgery is nitrous oxide, also known as laughing gas. Nitrous oxide is a colorless and odorless gas that has been used in medicine as an anesthetic since the mid-19th century. It is often used in combination with other anesthetic drugs, such as intravenous anesthetics, to provide more complete anesthesia.

Nitrous oxide works by inhibiting the transmission of nerve impulses and blocking the sensation of pain. It also has a relaxing effect on the body, which can help to reduce anxiety and induce a sense of well-being. In addition to its anesthetic properties, nitrous oxide has also been used as a pain reliever, particularly in dentistry.

Other anesthetic gases used in surgery include sevoflurane, desflurane, and isoflurane. These gases are all halogenated ethers, which means they contain halogen atoms, such as fluorine, chlorine, or bromine. Halogenated ethers are particularly effective as anesthetics because they are highly soluble in lipids and other fatty tissues, which allows them to quickly penetrate the blood-brain barrier and reach the brain.

Anesthetic gases are typically administered to patients through a face mask or an endotracheal tube, which is inserted into the patient's trachea. The concentration of the anesthetic gas is carefully monitored throughout the procedure to ensure that the patient remains in a safe and stable state of anesthesia.

While the use of anesthetic gases has greatly improved the safety and effectiveness of surgical procedures, they are not without risks. Some patients may experience side effects from the administration of anesthetic gases, such as nausea, vomiting, or allergic reactions. In rare cases, patients may also experience more serious complications, such as respiratory or cardiovascular problems.

To minimize the risk of complications, anesthesiologists carefully evaluate each patient before administering anesthesia and monitor their vital signs throughout the procedure. They also use a variety of techniques, such as regional anesthesia or conscious sedation, to tailor the anesthesia to the specific needs of each patient.

In conclusion, anesthetic gases are a critical component of modern surgical procedures. Nitrous oxide is the most commonly used anesthetic gas in surgery, but other halogenated ethers, such as sevoflurane, desflurane, and isoflurane, are also used. While anesthetic gases are generally safe and effective, they can cause side effects and complications, so their use must be carefully monitored and tailored to the individual needs of each patient.

38

What is heat stroke?

Heat stroke is a serious and potentially life-threatening condition that occurs when the body's core temperature rises to 104°F (40°C) or higher. It typically occurs as a result of prolonged exposure to high temperatures, often in combination with dehydration and physical exertion.

Heat stroke can affect anyone, but it is most common in older adults, young children, and people who work or exercise outdoors in hot and humid conditions. It can also occur in people who are confined to hot and poorly ventilated spaces, such as cars or buildings without air conditioning.

The symptoms of heat stroke typically develop gradually over several hours and may include:

- High body temperature (104°F or higher)
- Rapid pulse
- Rapid breathing
- Headache
- Dizziness or lightheadedness
- Nausea and vomiting
- Confusion, disorientation, or hallucinations
- Seizures
- Loss of consciousness

If left untreated, heat stroke can lead to serious complications, such as organ damage, brain damage, and even death.

Treatment for heat stroke involves rapidly lowering the body temperature using cooling techniques such as ice packs, cool water immersion, or evaporative cooling. In severe cases, hospitalization may be necessary to monitor vital signs and provide intravenous fluids and other medical interventions.

Prevention is key in avoiding heat stroke. It is important to stay hydrated, wear lightweight and loose-fitting clothing, and avoid prolonged exposure to hot and humid environments. It is also recommended to take frequent breaks in shaded or air-conditioned areas and to avoid strenuous activity during the hottest parts of the day.

In conclusion, heat stroke is a serious and potentially life-threatening condition that can occur in anyone exposed to high temperatures for prolonged periods of time. It is important to be aware of the symptoms and take steps to prevent heat stroke, particularly during the hot summer months or when working or exercising outdoors in hot and humid conditions.

39

What are the chemicals found in cigarette smoke?

Cigarette smoke contains a complex mixture of chemicals, including both toxic and carcinogenic compounds. Some of the main chemicals found in cigarette smoke are:

1. Nicotine: Nicotine is an addictive substance found in tobacco. It stimulates the nervous system and increases heart rate and blood pressure.
2. Tar: Tar is a mixture of many chemicals that are produced when tobacco is burned. It is a sticky brown substance that coats the lungs and can lead to lung cancer.
3. Carbon monoxide: Carbon monoxide is a poisonous gas that is produced when tobacco is burned. It reduces the amount of oxygen that can be carried in the blood, leading to a range of health problems.
4. Formaldehyde: Formaldehyde is a colorless gas with a pungent odor. It is a carcinogen that can cause cancer of the throat, nose, and lungs.
5. Benzene: Benzene is a colorless liquid that has a sweet odor. It is a carcinogen that can cause leukemia and other blood cancers.
6. Acrolein: Acrolein is a colorless liquid with a pungent odor. It is a toxic irritant that can cause damage to the lungs and airways.

7. Ammonia: Ammonia is a colorless gas with a pungent odor. It is used to enhance the flavor of tobacco and is a respiratory irritant.
8. Acetaldehyde: Acetaldehyde is a colorless liquid with a fruity odor. It is a carcinogen that can cause cancer of the throat, liver, and breast.
9. Nitrosamines: Nitrosamines are a group of carcinogenic chemicals that are produced during the tobacco curing process.

These chemicals are just a few of the many substances found in cigarette smoke. There are over 7,000 chemicals in cigarette smoke, with many of them being toxic or carcinogenic. It is important to note that the chemicals in cigarette smoke not only harm the smoker, but also those around them who inhale secondhand smoke.

40

How do refrigerators work?

Refrigerators have become an indispensable part of modern life, allowing us to store food and beverages at cool temperatures to keep them fresh for longer periods of time. But how do refrigerators work?

Refrigerators operate on the principle of heat transfer, which involves the transfer of heat energy from one place to another. In the case of a refrigerator, the heat is transferred from the inside of the refrigerator to the outside, which allows the interior of the refrigerator to remain cool.

The cooling process in a refrigerator is accomplished through a refrigerant, which is a chemical that has the ability to absorb and release heat energy. The refrigerant circulates through a series of pipes or coils in the refrigerator, absorbing heat energy as it passes through the interior of the refrigerator and releasing it as it passes through the exterior.

The refrigeration process begins in the compressor, which is typically located at the bottom of the refrigerator. The compressor circulates the refrigerant through the coils and pipes, compressing the gas and increasing its temperature. This high-pressure gas is then forced into the condenser, which is located at the back of the refrigerator.

In the condenser, the refrigerant is cooled down, causing it to condense into a liquid state. As the refrigerant changes from a gas to a liquid, it releases heat energy, which is then expelled from the refrigerator through the condenser coils.

The refrigerant then flows through the expansion valve, which is a small device that reduces the pressure of the refrigerant and allows it to expand back into a gas. As the refrigerant expands, it absorbs heat energy from the interior of the refrigerator, which cools the contents inside.

Finally, the refrigerant flows back into the compressor, where the cycle starts all over again.

In addition to the refrigerant, other components in the refrigerator also play important roles in the cooling process. The evaporator, for example, is a set of coils located in the interior of the refrigerator that help to absorb heat energy from the food and beverages stored inside. The thermostat, which is typically located in the refrigerator compartment, regulates the temperature by turning the compressor on and off as needed to maintain the desired temperature.

Overall, refrigerators use a combination of chemical and mechanical processes to keep our food and beverages fresh and cool, making them an essential part of modern life.

41

What is the role of zinc, potassium, and magnesium in our food?

Zinc, potassium, and magnesium are important nutrients that play essential roles in maintaining our health. They are found in various foods and are necessary for proper bodily functions.

Zinc is a trace mineral that is required for numerous cellular processes in the body. It is involved in the regulation of gene expression, cell growth and differentiation, and immune function. Zinc also plays a role in wound healing and the maintenance of skin and hair health. It is found in a variety of foods, including oysters, beef, pork, chicken, beans, nuts, and whole grains.

Potassium is an electrolyte that is important for maintaining fluid and electrolyte balance in the body. It is also involved in muscle contraction, nerve function, and the regulation of blood pressure. Potassium is found in many fruits and vegetables, including bananas, oranges, tomatoes, potatoes, and spinach.

Magnesium is a mineral that is involved in numerous biochemical reactions in the body. It is required for the synthesis of DNA, RNA, and proteins, and is involved in energy metabolism, muscle and nerve function, and the regulation of blood pressure. Magnesium is found in many foods, including green leafy vegetables, nuts, seeds, and whole grains.

A deficiency in any of these nutrients can lead to various health problems.

For example, a zinc deficiency can result in impaired immune function, delayed wound healing, and skin and hair problems. Potassium deficiency can lead to muscle weakness, cramping, and an irregular heartbeat. Magnesium deficiency can cause muscle twitches and cramps, fatigue, and mood changes.

It is important to include a variety of foods in our diet to ensure that we are getting adequate amounts of these essential nutrients. In some cases, supplements may be recommended for individuals who are at risk for deficiencies or who have medical conditions that affect nutrient absorption or utilization.

42

How is the age of our planet calculated, and what is the C14 method?

The age of our planet, Earth, has been a topic of scientific investigation for many centuries. Scientists have used various methods to determine the age of our planet, including radiometric dating techniques. One such technique is the Carbon-14 (C14) method, which is used to determine the age of organic materials.

The C14 method is based on the radioactive decay of carbon-14, a rare and unstable isotope of carbon that is formed in the upper atmosphere when cosmic rays collide with nitrogen atoms. Carbon-14 is absorbed by plants during photosynthesis and then enters the food chain when animals eat the plants. When an organism dies, the carbon-14 in its body begins to decay into nitrogen-14 at a known rate, with a half-life of approximately 5,700 years.

By measuring the ratio of carbon-14 to carbon-12 in an organic material, scientists can determine the amount of carbon-14 that has decayed and estimate the age of the material. The C14 method is limited to materials that are less than 50,000 years old, as the amount of carbon-14 left in the material after this time period is too small to measure accurately.

Another method used to determine the age of our planet is through the study of rocks and minerals. This method is called radiometric dating, which uses the decay of radioactive isotopes to determine the age of rocks and minerals.

Uranium-lead dating is a commonly used radiometric dating technique, which involves the decay of uranium-238 into lead-206. The age of rocks and minerals can be estimated by measuring the ratio of uranium-238 to lead-206 in the sample.

Other methods used to determine the age of the Earth include the study of the Earth's magnetic field and the rate of cooling of the Earth's core. These methods provide additional data that can be used to estimate the age of the Earth.

The estimated age of the Earth using various methods is approximately 4.54 billion years old. This estimate is based on a combination of radiometric dating techniques, the study of rocks and minerals, and other methods used to determine the age of the Earth. The C14 method is just one of many techniques used to estimate the age of the Earth, and it is a powerful tool for studying the history of life on our planet.

43

Accidentally discovered substances

Throughout history, many substances have been accidentally discovered. These substances range from common household items to life-saving medications. In this article, we will explore some of the most notable examples of accidentally discovered substances and their impact on society.

Penicillin

One of the most well-known examples of an accidentally discovered substance is penicillin. In 1928, Scottish biologist Alexander Fleming was conducting research on the flu virus when he noticed a mold growing on one of his Petri dishes. He discovered that the mold was killing the bacteria on the dish, which led him to further investigate the properties of the mold. He identified the mold as Penicillium notatum and discovered that it produced a substance that had antibacterial properties. This substance, which he named penicillin, went on to revolutionize the field of medicine by providing an effective treatment for bacterial infections.

Saccharin

Saccharin is a popular artificial sweetener that was discovered by accident in the late 1800s. A chemist named Constantin Fahlberg was working on a coal tar derivative when he accidentally spilled some of the substance on his hands. Later that evening, he noticed that the bread he was eating tasted sweet. He realized that the sweet taste came from the chemical he had spilled on his hands. He went on to develop saccharin as a sweetener, which is still used today in a variety of products.

Teflon

Teflon is a non-stick coating used in a variety of cookware and other products. It was discovered accidentally in 1938 by chemist Roy Plunkett while he was working on developing a new refrigerant. Plunkett discovered that the gas he was working with had solidified inside the container, which led him to investigate further. He found that the gas had formed a slippery coating on the inside of the container, which he identified as Teflon. Teflon has since become a widely used material due to its non-stick properties.

Super Glue

Super Glue is a popular adhesive that was discovered by accident in 1942 by a team of scientists working for Kodak. They were attempting to develop a clear plastic that could be used for gun sights during World War II. However, they ended up creating a sticky substance that was difficult to work with. Several years later, a chemist named Harry Coover rediscovered the substance and realized its potential as an adhesive. Super Glue has since become a household name and is used in a variety of applications.

Viagra

Viagra is a medication used to treat erectile dysfunction. It was discovered accidentally in the 1990s by a team of researchers at Pfizer who were working on a medication to treat high blood pressure and angina. During clinical trials, they found that the medication had an unexpected side effect - it caused erections in male patients. This led to further investigation and the development of Viagra as a treatment for erectile dysfunction.

In conclusion, these are just a few examples of the many accidentally discovered substances that have had a profound impact on society. From life-saving medications to common household products, these discoveries serve as a reminder that sometimes the most significant breakthroughs can come from unexpected places.

44

Does a mixture of two liters of water and two liters of alcohol make four liters?

No, a mixture of two liters of water and two liters of alcohol does not make four liters. This is because the mixture undergoes a process called contraction, where the total volume of the mixture is less than the sum of the volumes of the individual components.

This is because water molecules are more attracted to each other than to alcohol molecules, and similarly, alcohol molecules are more attracted to each other than to water molecules. When the two are mixed together, the water molecules will form clusters and the alcohol molecules will form clusters, reducing the overall volume of the mixture.

The degree of contraction depends on a number of factors, including the types of molecules in the mixture and their relative concentrations. In general, the more dissimilar the molecules and the more different their concentrations, the greater the degree of contraction.

So, in the case of a mixture of two liters of water and two liters of alcohol, the total volume of the mixture will be less than four liters. The exact amount of contraction will depend on the specific properties of the water and alcohol being used.

45

Does tea with sugar or without sugar cool faster?

When it comes to cooling a hot beverage like tea, adding sugar or not can make a difference in the rate at which the tea cools down.

First, let's look at the science behind how heat is transferred. Heat transfer occurs in three ways: conduction, convection, and radiation. Conduction is the transfer of heat through a solid or stationary medium, such as a pot on a stove. Convection is the transfer of heat through a fluid or gas, such as hot air rising from a stove. Radiation is the transfer of heat through electromagnetic waves, such as the heat felt from the sun.

Now, let's consider how sugar affects the cooling rate of tea. Adding sugar to tea increases the overall mass of the solution, which means that it will take more energy to cool down. However, sugar also makes the solution more viscous, which can reduce the rate of convection. This means that the tea with sugar will cool down slower than tea without sugar.

On the other hand, tea without sugar has a lower overall mass and is less viscous than tea with sugar. This means that the tea without sugar will cool down faster than tea with sugar.

It's important to note that the difference in cooling rate between tea with sugar and without sugar is likely to be small and may not be noticeable in practical situations. Additionally, other factors such as the size and shape of the container, the initial temperature of the tea, and the surrounding temperature can also affect the cooling rate.

In summary, tea with sugar will cool down slower than tea without sugar due to its higher overall mass and increased viscosity. However, the difference in cooling rate may not be significant in practical situations.

46

How much energy is released during the explosion of an atom or hydrogen bomb?

The explosion of an atom or hydrogen bomb is a catastrophic event that releases an enormous amount of energy. This energy is the result of a process called nuclear fission, in which the nucleus of an atom is split into two smaller nuclei, releasing a vast amount of energy in the process.

The amount of energy released during a nuclear explosion is difficult to measure precisely because it depends on many factors, including the size of the bomb, the type of nuclear material used, and the conditions under which the explosion occurs. However, it is generally accepted that the energy released in a single atomic bomb explosion can range from a few kilotons (thousands of tons) of TNT to several megatons (millions of tons) of TNT.

For example, the atomic bomb dropped on Hiroshima during World War II, which had a yield of approximately 15 kilotons of TNT, released an enormous amount of energy, equivalent to the explosion of 15,000 tons of TNT. The energy released by the more powerful hydrogen bombs can be even more staggering. The most powerful nuclear device ever tested, the Tsar Bomba, had a yield of 50 megatons of TNT, or the equivalent of the explosion of 50 million tons of TNT.

The energy released during a nuclear explosion is primarily in the form of thermal radiation, which is the same type of radiation that causes sunburn.

This radiation is so intense that it can cause severe burns and even ignite combustible materials many miles away from the explosion.

In addition to the immediate effects of the explosion, the long-term effects of radiation exposure can be devastating. The radiation can cause cancer, genetic mutations, and other serious health problems that can persist for generations.

The energy released during a nuclear explosion is so vast that it can have a significant impact on the environment as well. The explosion can cause massive fires, generate a shock wave that can destroy buildings and other structures, and release large amounts of radioactive material into the atmosphere, which can spread over a wide area and contaminate the environment.

In conclusion, the energy released during the explosion of an atom or hydrogen bomb is vast and devastating. The destructive force of such an explosion can cause massive damage to both human life and the environment. It is crucial to prevent the proliferation of nuclear weapons and work towards nuclear disarmament to ensure a safer and more peaceful world.

47

What is the internal structure of inks?

Inks are a mixture of various compounds that are used to produce writing or printing on surfaces such as paper, fabric, or skin. The composition of inks varies depending on their intended use, such as whether they are used for writing or printing, the type of surface they are used on, and the desired color and consistency.

The internal structure of inks is quite complex and can be broken down into several components. The main components of inks are pigments, binders, solvents, and additives.

Pigments are the colorants used in inks. They are finely ground particles that give the ink its color. Pigments can be organic or inorganic, and their chemical structure determines their properties. Inorganic pigments are often made of metals or metal oxides, while organic pigments are typically made of carbon-based compounds.

Binders are compounds that hold the pigment particles together and allow them to adhere to the surface. They are also responsible for controlling the ink's viscosity and drying time. Common binders include resins, polymers, and waxes.

Solvents are liquids that dissolve the pigments and binders, allowing them to be spread evenly across the surface. The type of solvent used depends on the type of ink and the surface it is being applied to. For example, water-based inks use water as a solvent, while oil-based inks use oil.

Additives are compounds added to inks to modify their properties. These may include surfactants to improve wetting, defoamers to reduce bubbles, and preservatives to prevent the growth of microorganisms.

The internal structure of inks can also vary depending on their method of application. For example, ballpoint pen inks contain a volatile solvent that evaporates quickly, while printer inks use a non-volatile solvent that does not evaporate as quickly.

In conclusion, the internal structure of inks is complex and consists of various components that work together to produce a desired color and consistency on a surface. Understanding the composition of inks is important in developing new ink formulations and improving their performance.

48

Why is mercury a liquid metal?

Mercury is a unique and fascinating element due to its unusual physical properties, particularly its liquid state at room temperature. Unlike most metals that are solid at room temperature, mercury is a liquid metal with a silvery appearance. So, why is mercury a liquid metal?

The reason behind mercury's liquid state is due to its electronic configuration and bonding. Mercury is a transition metal and belongs to Group 12 of the periodic table. It has an atomic number of 80 and its electronic configuration is [Xe] $4f14\ 5d10\ 6s2$. Mercury has a relatively low melting point of $-38.83\ °C$ and a boiling point of $356.73\ °C$.

The low melting point of mercury is due to its weak metallic bonding. In metallic bonding, the metal atoms are packed closely together, and their outermost valence electrons are delocalized, forming a sea of electrons that can move freely throughout the metal structure. This sea of electrons is responsible for the characteristic properties of metals, such as high electrical conductivity, malleability, and ductility.

However, in mercury, the $5d10$ electrons are tightly bound to the atom, and there are no electrons in the $6p$ orbitals that can participate in metallic bonding. Therefore, the bonding between mercury atoms is much weaker than other metals, resulting in a low melting point.

Additionally, the small size of mercury atoms also contributes to its liquid state at room temperature. Mercury atoms are small and have a low atomic radius, which results in weak van der Waals forces between the atoms. As a result, the energy required to overcome these forces and convert mercury from a liquid to a solid is relatively low.

The unique properties of mercury have made it useful in many applications, such as thermometers, barometers, electrical switches, and fluorescent lamps. However, it is also a toxic substance and can cause serious health problems if ingested or inhaled. Therefore, it is essential to handle mercury with care and take appropriate safety precautions when working with this element.

49

Is glass a liquid or a solid?

Glass is often referred to as an amorphous solid or a "supercooled liquid" rather than a true solid or liquid. The reason for this is that the atoms or molecules in glass do not form a crystalline structure like those in a true solid, but instead, they are arranged in a disordered and random fashion.

In a true solid, the atoms or molecules are arranged in an orderly fashion, forming a crystalline lattice structure. When heated, a solid will typically melt and become a liquid with a less ordered arrangement of molecules. When cooled, the liquid will typically solidify again, forming a crystalline structure.

However, with glass, the molecules are not arranged in a crystalline structure, but instead, they are frozen in a disordered and random state, similar to a liquid but without the ability to flow like a liquid. This is because glass is formed by rapidly cooling a liquid to prevent the molecules from forming a crystalline structure. As a result, glass has some properties of both a solid and a liquid, but it is not exactly either one.

In summary, glass is not a true solid or liquid, but rather an amorphous solid that falls somewhere in between the two.

50

Why do our eyes water when we peel onions?

Have you ever wondered why your eyes start to tear up when you chop an onion? The answer lies in the chemistry of the onion itself.

Onions contain a molecule called syn-propanethial-S-oxide, or simply onion lachrymatory factor (LF). When you cut or chop an onion, you rupture the cells and release LF into the air. The LF then reacts with the moisture in your eyes to form a mild sulfuric acid. This acid irritates the nerve endings in your eyes and causes them to tear up as a reflex action.

But why do onions contain this molecule in the first place? It is actually a defense mechanism against predators. Onions are part of the allium family, which also includes garlic, shallots, and chives. All of these plants produce LF as a way to deter animals from eating them. LF is also responsible for the distinctive flavor and aroma of onions.

So, how can you prevent tearing up when chopping onions? One way is to chill the onion in the refrigerator for about 30 minutes before cutting it. This will slow down the release of LF and reduce the amount that reaches your eyes. You can also try cutting the onion under running water or wearing goggles to protect your eyes.

In conclusion, the tears that come with cutting onions are a result of the chemistry of the onion itself. While it may be an inconvenience, it is a natural defense mechanism that has evolved over time to protect the onion from being eaten by predators.

51

How do rechargeable batteries work?

Rechargeable batteries have become an essential component of modern-day life, powering devices ranging from smartphones and laptops to electric cars and power grids. These batteries store electrical energy in chemical form and can be recharged multiple times, making them a more sustainable and cost-effective alternative to single-use batteries. But how exactly do they work?

Rechargeable batteries come in many different types, including nickel-cadmium (NiCad), nickel-metal hydride (NiMH), and lithium-ion (Li-ion) batteries. While they all have slightly different chemistries, they all function based on the same basic principles.

At the heart of a rechargeable battery is a chemical reaction that produces electrons. When the battery is discharging (i.e., powering a device), this chemical reaction is exothermic, meaning it releases energy in the form of electrons. The electrons flow through a circuit to power the device, while the reaction products (i.e., the depleted chemicals) remain in the battery.

When the battery is recharged, the chemical reaction is reversed. An external power source is used to force electrons back into the battery, which reverses the chemical reaction and restores the battery's energy storage capacity. This process is not 100% efficient, however, and some energy is lost as heat during charging.

The specific chemistry of a rechargeable battery depends on its type. NiCad batteries, for example, use a nickel hydroxide cathode and a cadmium anode. During discharge, the cadmium is oxidized to cadmium hydroxide, and the nickel hydroxide is reduced to nickel oxyhydroxide. During charging, the process is reversed, and the nickel oxyhydroxide is oxidized back to nickel hydroxide while the cadmium hydroxide is reduced back to cadmium.

NiMH batteries have a similar chemistry, but they use a hydrogen-absorbing alloy instead of cadmium as the anode material. This allows them to store more energy than NiCad batteries of the same size and weight.

Li-ion batteries, which are commonly used in smartphones and laptops, use a lithium cobalt oxide cathode and a graphite anode. During discharge, the lithium ions move from the cathode to the anode, while electrons flow through the external circuit to power the device. During charging, the lithium ions are forced back into the cathode, where they are stored until the next discharge cycle.

One of the key advantages of rechargeable batteries is their ability to be recharged multiple times. However, this ability is not infinite, and rechargeable batteries will eventually degrade and lose their ability to hold a charge. This can happen due to a number of factors, including exposure to heat, overcharging, and simply using the battery over time.

In summary, rechargeable batteries are an essential component of modern-day life, and they work by storing energy in chemical form and releasing it as electrical energy when needed. While they all have slightly different chemistries, they all function based on the same basic principles of electron flow and chemical reactions. By understanding how rechargeable batteries work, we can better appreciate their importance in our daily lives and work towards improving their efficiency and sustainability.

52

Is fire matter or energy?

Fire is the visible result of a chemical reaction between oxygen and a fuel source, which produces heat and light. In this sense, fire is both matter and energy, as it is the product of a chemical reaction (matter) and releases energy in the form of heat and light.

The chemistry of fire involves a process known as combustion, which is a reaction between a fuel source and oxygen, typically from the air. The fuel source can be any material that can burn, including wood, coal, oil, gasoline, and natural gas, among others. When these materials are heated to their ignition temperature, they react with oxygen in the air, producing a flame and releasing heat and light.

The chemical reaction that produces fire is an exothermic reaction, which means that it releases heat. The heat generated by the reaction causes the surrounding air to heat up and expand, creating a convection current that draws more oxygen into the flame and feeds the reaction. This creates a self-sustaining chain reaction, which is why fires can continue to burn as long as there is fuel and oxygen available.

The color of the flame in a fire depends on the temperature of the flame and the chemical composition of the fuel source. Blue flames, for example, are typically hotter than yellow or orange flames, and are often produced by fuels that contain higher concentrations of carbon, such as natural gas. Yellow or orange flames, on the other hand, are cooler and are typically produced by fuels that contain lower concentrations of carbon, such as wood.

In addition to producing heat and light, fires can also release a variety of gases and particulate matter, including carbon dioxide, carbon monoxide, and smoke. These emissions can be harmful to human health and the environment, which is why it is important to take steps to prevent and control fires, and to use fuels and materials that minimize emissions.

In conclusion, fire is both matter and energy, as it is the result of a chemical reaction between a fuel source and oxygen, and releases heat and light in the process. The chemistry of fire is complex, and involves a self-sustaining chain reaction that produces a variety of

53

What is the chemical structure of butter?

Butter is a dairy product made from milk or cream. The chemistry of butter is quite complex, but its main components are fats, water, and some minor components that give it its unique flavor and aroma.

The primary fats in butter are triglycerides, which are composed of three fatty acid molecules attached to a glycerol molecule. The type of fatty acid can vary depending on the animal and diet of the animal that the milk comes from, but in general, butter contains a mixture of saturated, monounsaturated, and polyunsaturated fatty acids.

One of the most abundant fatty acids in butter is called palmitic acid, which is a saturated fatty acid that makes up about 25-30% of the total fatty acids. Another common saturated fatty acid in butter is stearic acid, which makes up about 12-16% of the total fatty acids. Oleic acid is a monounsaturated fatty acid that makes up about 20-30% of the total fatty acids, and linoleic and linolenic acids are polyunsaturated fatty acids that make up about 2-4% and 0.5-1% of the total fatty acids, respectively.

The water content of butter is also important and can vary depending on the type of butter. Typically, butter contains around 15-20% water, but some types of butter, such as cultured butter, can contain up to 30% water.

In addition to fats and water, butter also contains minor components such as milk proteins, vitamins, minerals, and compounds that give it its characteristic flavor and aroma. For example, butter contains small amounts of the fat-soluble vitamins A, D, E, and K, as well as the water-soluble vitamin B12. Butter also contains minerals such as calcium, phosphorus, and potassium.

Butter also contains compounds that give it its characteristic flavor and aroma, such as diacetyl, which has a buttery flavor, and 2,3-pentanedione, which has a caramel-like aroma. These compounds are formed during the process of fermentation, which occurs when bacteria in the milk convert lactose into lactic acid.

In conclusion, the chemistry of butter is quite complex, but it is primarily composed of fats, water, and minor components such as milk proteins, vitamins, minerals, and compounds that give it its characteristic flavor and aroma. The composition of butter can vary depending on the type of butter and the animal and diet of the animal that the milk comes from.

54

What is the chemical structure of lipstick?

Lipstick is a cosmetic product that is used to add color and texture to the lips. It is made up of a variety of different ingredients, including waxes, oils, pigments, and emollients. The exact chemical composition of lipstick can vary depending on the brand, formulation, and color.

The main ingredient in most lipsticks is a mixture of waxes, including beeswax, carnauba wax, and candelilla wax. These waxes provide the structure and stability to the product, helping it to stay on the lips and maintain its shape. In addition, the waxes can help to create a smooth, even surface on the lips and provide a protective barrier against moisture loss.

Oils are also a key component of lipstick formulations. They help to provide moisture to the lips, preventing them from becoming dry and chapped. Common oils used in lipstick include castor oil, jojoba oil, and mineral oil.

To create the color of the lipstick, pigments are added to the formulation. These pigments can be derived from a variety of sources, including minerals, plants, and synthetic dyes. Some common pigments used in lipstick include iron oxides, titanium dioxide, and carmine.

Emollients are another important ingredient in lipstick formulations. They help to soften and smooth the lips, providing a comfortable texture and helping the lipstick to glide on smoothly. Common emollients used in lipstick include lanolin, shea butter, and cocoa butter.

Other ingredients that may be found in lipstick formulations include fragrances, preservatives, and antioxidants. These ingredients can help to improve the scent and shelf life of the product, as well as protect it from oxidation and other forms of damage.

Overall, the chemical structure of lipstick is complex and can vary depending on the specific formulation. However, by understanding the basic ingredients and their functions, consumers can make more informed choices when selecting a lipstick product that meets their needs and preferences.

55

Why should we drink carbonated beverages like cola cold?

Carbonated beverages like cola are enjoyed by many people around the world for their refreshing and bubbly taste. But have you ever wondered why they taste best when cold? The answer lies in the chemistry of carbon dioxide (CO_2) gas, which is responsible for making these drinks fizzy.

When a bottle or can of cola is opened, the pressure inside the container decreases, allowing some of the dissolved CO_2 to escape in the form of bubbles. This is because the solubility of gases in liquids decreases as the temperature increases, and so warmer cola can hold less CO_2 than colder cola.

So, if you were to drink warm cola, the bubbles would escape more quickly and the drink would go flat much faster. By contrast, when you drink cold cola, the CO_2 stays dissolved in the liquid for longer, keeping the drink fizzy and refreshing for longer.

But the relationship between temperature and carbonation is more complex than simply a matter of solubility. When a liquid is cold, its viscosity increases, meaning that it becomes thicker and more syrupy. This has the effect of slowing down the release of bubbles from the liquid, so even if the solubility of CO_2 were the same in cold and warm cola, the drink would still be fizzier when it is cold.

Another reason why carbonated beverages should be drunk cold is that the taste of cold drinks is generally perceived as more intense and refreshing than warm drinks. This is due to the fact that colder temperatures reduce our ability to taste sweetness and bitterness, while enhancing our perception of acidity and carbonation.

In conclusion, the chemistry of carbonation and temperature plays a significant role in the taste and texture of carbonated beverages like cola. Drinking these drinks cold is essential for maintaining their fizziness and refreshing taste, as well as enhancing our perception of their flavors.

56

Can we make diamonds from graphite?

Diamonds are one of the most precious and valuable gemstones on Earth. They are known for their remarkable physical and optical properties, such as their extreme hardness, high refractive index, and beautiful sparkle. Diamonds are made of carbon atoms arranged in a crystal lattice structure, and they can form deep within the Earth's mantle under conditions of high temperature and pressure. But can we make diamonds from graphite, which is also made of carbon atoms but has a different crystal structure? The answer is yes, it is possible to turn graphite into diamonds using high pressure and high temperature.

Graphite and diamond are both allotropes of carbon, meaning they are different forms of the same element. Graphite is a soft, black, opaque substance used in pencils, lubricants, and batteries, among other things. Its crystal structure consists of layers of carbon atoms arranged in hexagons, similar to a honeycomb pattern. The layers are held together by weak intermolecular forces, which allow them to slide over each other, making graphite a good lubricant.

Diamond, on the other hand, is a clear, hard, and transparent substance used in jewelry, cutting tools, and electronics, among other things. Its crystal structure consists of carbon atoms arranged in a three-dimensional lattice structure, similar to a pyramid pattern. The atoms are held together by strong covalent bonds, which make diamond extremely hard and durable.

So how do we turn graphite into diamond? The process is called high-pressure high-temperature (HPHT) synthesis, and it involves subjecting graphite to extreme conditions of pressure and temperature. The process is similar to how diamonds form naturally in the Earth's mantle, except that it is done in a laboratory using specialized equipment.

During the HPHT process, graphite is placed in a diamond anvil cell, which is a small device that can create extremely high pressure by squeezing the sample between two diamond anvils. The pressure can be as high as several million pounds per square inch, which is equivalent to the pressure deep within the Earth's mantle. The cell is then heated to a high temperature using a laser or other heat source. The combination of high pressure and high temperature causes the carbon atoms in the graphite to rearrange into a diamond lattice structure.

The resulting diamonds are usually small and of low quality, but they are genuine diamonds nonetheless. They can be used for industrial purposes, such as abrasives or cutting tools, but they are not suitable for jewelry because of their small size and low quality. However, the HPHT process has opened up new possibilities for diamond synthesis and has led to advances in the field of materials science and engineering.

In conclusion, it is possible to turn graphite into diamonds using high-pressure high-temperature synthesis. The process involves subjecting graphite to extreme conditions of pressure and temperature, causing the carbon atoms to rearrange into a diamond lattice structure. While the resulting diamonds are of low quality, the HPHT process has opened up new possibilities for diamond synthesis and has led to advances in materials science and engineering.

57

Why is silicon used in computer chip production?

Silicon is used in computer chip production because of its semiconductor properties. A semiconductor is a material that conducts electricity under certain conditions but not others, and silicon is a particularly good semiconductor. Silicon has four valence electrons, which means that it can easily form covalent bonds with other silicon atoms to create a crystal lattice structure. However, these bonds are not strong enough to completely hold onto their electrons, which allows the electrons to move around and conduct electricity.

Computer chips are made up of millions of tiny transistors, which are essentially switches that turn on and off to represent the 0's and 1's of binary code. These transistors are made using a process called photolithography, which involves layering and patterning thin films of silicon and other materials onto a silicon wafer. The wafer is then etched and doped with impurities to create the desired electrical properties.

Silicon is used in this process because it is abundant, easy to purify, and has a high melting point, making it a durable material for manufacturing. Additionally, its semiconducting properties can be easily manipulated by adding impurities, or dopants, to create different types of electrical conductivity. By controlling the doping levels, manufacturers can create p-type and n-type semiconductors, which can be combined to create the p-n junctions that make up transistors.

Overall, silicon's unique properties make it an ideal material for the production of computer chips and other electronic devices.

58

What is a laser?

A laser is a device that emits a beam of light through a process called stimulated emission. The term "laser" is an acronym that stands for "Light Amplification by Stimulated Emission of Radiation."

The basic structure of a laser consists of a gain medium, which is typically made of a material such as a crystal, gas, or semiconductor. The gain medium is excited by an external energy source, such as an electrical current or a flash of light. This excitation causes the gain medium to emit photons, or particles of light, which are then amplified as they bounce back and forth through a cavity between two mirrors.

The photons that are emitted by the gain medium have the same wavelength and are in phase with each other, which means they are all moving in the same direction and are aligned in their peaks and troughs. This results in a coherent beam of light that is highly directional and can travel over long distances without losing its intensity.

Lasers are used in a wide range of applications, including scientific research, communication, medicine, manufacturing, and entertainment. Some common examples of laser applications include laser pointers, barcode scanners, laser cutting and welding, laser surgery, and laser light shows.

59

Can the percentage of alcohol obtained by fermentation exceed 14%?

Alcoholic fermentation is a natural process in which yeast breaks down sugar into alcohol and carbon dioxide. However, the maximum alcohol percentage that can be obtained by fermentation is typically around 14%, because beyond this point, the alcohol begins to kill off the yeast responsible for the fermentation process.

There are several factors that can affect the maximum alcohol percentage obtained through fermentation, including the type of yeast used, the temperature at which the fermentation occurs, and the amount of sugar present in the initial mixture. In some cases, specialized strains of yeast can be used that are capable of tolerating higher levels of alcohol, allowing for a slightly higher alcohol percentage to be obtained.

However, it is generally not possible to obtain an alcohol percentage greater than around 20% through fermentation alone. For higher alcohol percentages, additional distillation processes are typically required.

60

How does the size of atoms change in the periodic table and what are the exceptions?

The size of atoms generally decreases from left to right across a period in the periodic table, and increases from top to bottom down a group. This trend is due to the increasing number of protons in the nucleus and the increasing number of electron shells.

As the number of protons in the nucleus increases, the attractive force between the positively charged nucleus and negatively charged electrons increases. This results in a decrease in the atomic radius.

On the other hand, as we move down a group, the number of electron shells increases, which results in an increase in the atomic radius. This is because the outermost electrons are farther away from the positively charged nucleus and are shielded by the inner electrons.

There are, however, some exceptions to this trend. For example, the atomic radius of helium is smaller than that of hydrogen, despite being in the same period. This is because helium has a full electron shell, which makes it more stable than hydrogen.

Another exception is that the atomic radius of the transition metals in the middle of the periodic table is relatively constant. This is due to the fact that the electrons in the d subshell are not very effective at shielding the outermost electrons from the positively charged nucleus.

In summary, the size of atoms in the periodic table generally follows a pattern, but there are some exceptions due to factors such as electron configuration and the effectiveness of electron shielding.

61

What are the metals and gases inside a light bulb?

A light bulb is a device that produces light by passing an electric current through a filament or gas. The two most common types of light bulbs are incandescent bulbs and fluorescent bulbs.

Incandescent bulbs contain a filament made of tungsten, a metal that is known for its high melting point and ability to withstand high temperatures. The filament is enclosed in a glass bulb that is filled with an inert gas, usually argon or nitrogen, to prevent the filament from oxidizing and burning out.

When an electric current is passed through the filament, it heats up and emits light. The color of the light depends on the temperature of the filament.

Fluorescent bulbs, on the other hand, contain a gas mixture that is enclosed in a glass tube. The gas mixture usually contains argon, neon, and a small amount of mercury vapor. When an electric current is passed through the gas, it excites the mercury atoms, causing them to emit ultraviolet radiation. The ultraviolet radiation then interacts with a phosphor coating on the inside of the tube, causing it to emit visible light.

In summary, the main metals found in a light bulb are tungsten, used as a filament in incandescent bulbs, and mercury, used in fluorescent bulbs. The gases found inside a light bulb are typically inert gases such as argon or nitrogen, used to prevent oxidation of the filament in incandescent bulbs, and a mixture of argon, neon, and mercury vapor in fluorescent bulbs.

62

What are the most abundant elements on Earth?

The Earth is made up of a variety of elements, but some are more abundant than others. The most abundant elements on Earth are oxygen, silicon, aluminum, iron, calcium, sodium, potassium, and magnesium. Together, they make up more than 98% of the Earth's crust by weight.

Oxygen is the most abundant element on Earth, making up about 46% of the Earth's crust by weight. It is also the most abundant element in the Earth's atmosphere, making up about 21% of the air we breathe.

Silicon is the second most abundant element on Earth, making up about 28% of the Earth's crust by weight. It is an essential component of many minerals and rocks, including quartz, feldspar, and mica.

Aluminum is the third most abundant element on Earth, making up about 8% of the Earth's crust by weight. It is a lightweight, durable metal that is used in a wide range of applications, including construction, transportation, and packaging.

Iron is the fourth most abundant element on Earth, making up about 5% of the Earth's crust by weight. It is a strong, durable metal that is used in many industrial and construction applications, as well as in the production of steel.

Calcium, sodium, potassium, and magnesium are all important minerals that are essential for human health. Calcium is the fifth most abundant element on Earth, making up about 3% of the Earth's crust by weight. Sodium and potassium are both alkali metals that are important for regulating fluid balance and nerve function in the body. Magnesium is an essential mineral that is involved in many biological processes, including energy production and protein synthesis.

While these elements are the most abundant on Earth, there are many other elements that are also important for life and for the functioning of the Earth's ecosystem.

63

Why is a lithium battery more advantageous?

Lithium batteries are a type of rechargeable battery that are commonly used in a variety of electronic devices, including cell phones, laptops, and electric vehicles. They are more advantageous than other types of batteries for several reasons:

1. High energy density: Lithium batteries have a high energy density, which means they can store a lot of energy in a relatively small amount of space. This makes them ideal for use in portable electronic devices, where space is at a premium.
2. Long cycle life: Lithium batteries can be recharged and discharged many times without losing their capacity, which means they can last for years without needing to be replaced.
3. Low self-discharge rate: Lithium batteries have a low self-discharge rate, which means they can hold their charge for a long time even when not in use.
4. High power output: Lithium batteries can deliver a high power output, which makes them suitable for use in high-performance applications such as electric vehicles.
5. Lightweight: Lithium batteries are lightweight, which makes them ideal

for use in portable electronic devices.

6. Environmentally friendly: Lithium batteries are more environmentally friendly than other types of batteries because they contain less toxic metals and chemicals.

Overall, the advantages of lithium batteries make them an attractive option for many different applications, from portable electronic devices to electric vehicles.

64

Are you magnetic?

Humans are not inherently magnetic in the sense of having a magnetic field that can be detected by other magnets or magnetic sensors. However, the human body does produce tiny electrical currents that generate a magnetic field, albeit a very weak one. This field, known as the biomagnetic field or the human magnetic field, is created by the activity of electrically charged particles in the body, such as nerve impulses and muscle contractions.

The strength of the human magnetic field is very small, on the order of a few millionths of a tesla (a unit of magnetic field strength). By comparison, the Earth's magnetic field has a strength of about 50 microtesla at its surface. The human magnetic field is also highly localized, meaning that it varies in strength and direction depending on the location and activity of the body part in question.

Despite their weak magnetic fields, some researchers believe that humans may be able to detect magnetic fields, much like some animals such as birds and fish. However, the evidence for this ability in humans is still controversial and inconclusive.

It is worth noting that some people may have magnetic implants, such as those used in magnetic therapy or in certain medical procedures. These implants can create a magnetic field that can interact with external magnets or magnetic fields. However, this is not a natural or inherent property of the human body.

65

Is your water hard or soft?

Hard water and soft water are two terms used to describe the mineral content in water. Hard water is water that has a high concentration of dissolved minerals, such as calcium and magnesium. Soft water, on the other hand, has a low concentration of these minerals.

The hardness of water can vary depending on the location and source of the water. Water that comes from underground sources, such as wells, is often harder than water that comes from surface sources, such as rivers and lakes.

There are several ways to determine if your water is hard or soft. One way is to look for the buildup of white or gray deposits on your faucets, showerheads, and other plumbing fixtures. This buildup is caused by the minerals in hard water.

Another way to determine water hardness is to perform a simple soap test. Soft water lathers easily with soap, while hard water does not. To perform the test, place a small amount of soap in a container with water and shake it vigorously. If the water lathers easily, it is soft. If the water does not lather or forms a scum, it is hard.

Hard water can have several negative effects. It can leave stains on clothing and dishes, clog pipes, and reduce the effectiveness of soaps and detergents. It can also make hair and skin feel dry and itchy.

To soften hard water, a water softener can be installed in your home. A water softener is a device that removes the minerals that cause hardness from the water. It uses a process called ion exchange, where the hard minerals are exchanged for sodium or potassium ions.

In conclusion, hard water contains a high concentration of minerals, while soft water contains a low concentration of minerals. There are several ways to determine water hardness, and hard water can have several negative effects. Water softeners can be installed to remove the minerals that cause hardness from the water.

66

How can you make the most of your soap?

It mentions that soap contains sodium stearate ($C_{17}H_{35}COONa$), which has a long hydrocarbon chain and is apolar, meaning it does not mix well with polar substances like water. However, soap also contains carboxyl ($COO-$) groups, which are ionic and like to react with water. When we use soap to wash our hands, the hydrocarbon chains in the soap absorb the dirt and oil on our skin, and the ionic carboxyl groups react with the water to wash it away. This is the basic principle behind the cleaning properties of soap, which also applies to other cleaning products like shampoo, toothpaste, and detergent.

Soap is a ubiquitous household product that we all use to cleanse ourselves, our clothes, and even our homes. But did you know that there are ways to make the most of your soap? Here are some tips and tricks to help you get the most out of your soap:

1. Store your soap properly: To make your soap last longer, it's important to store it properly. After use, make sure to let it dry out completely before storing it in a soap dish. This will prevent it from becoming soft and mushy.
2. Cut your soap into smaller pieces: Cutting your soap into smaller pieces can help make it last longer. This is especially useful if you have a large bar of soap that's difficult to hold onto.
3. Use a soap saver: A soap saver is a small pouch made of mesh or other

materials that helps to extend the life of your soap. Simply place your soap inside the pouch and use it to lather up.

4. Use a washcloth or loofah: Using a washcloth or loofah can help to create a rich lather with your soap and help it last longer. It also helps to exfoliate your skin, leaving it soft and smooth.

5. Use soap scraps: Don't throw away those small pieces of soap that are too small to use on their own. Instead, save them and use them to create a new bar of soap.

6. Use soap nuts: Soap nuts are a natural alternative to traditional soap. They are made from the fruit of the soapberry tree and can be used to create a natural cleaning solution for your laundry, dishes, and even your hair.

In summary, there are many ways to make the most of your soap. By storing it properly, cutting it into smaller pieces, using a soap saver, using a washcloth or loofah, using soap scraps, and trying out soap nuts, you can extend the life of your soap and get the most out of this essential household product.

67

Do you have a pillow in your car?

Airbags are an important safety feature in modern cars. They are designed to inflate rapidly in the event of a collision, cushioning the occupants and reducing the risk of serious injury. The gas that fills the airbag is usually nitrogen, which is generated by a chemical reaction that takes place inside the airbag module.

The chemical responsible for this reaction is sodium azide (NaN_3), a compound that is typically found in the airbag assembly. When a car collides with an object, a sensor detects the impact and sends a signal to the inflator module. This module contains a small amount of iron oxide (Fe_2O_3), which acts as a catalyst for the reaction.

When the signal is received, an electrical charge ignites a small amount of the iron oxide, which then reacts with the sodium azide to produce nitrogen gas (N_2) and solid sodium metal (Na). The reaction is extremely fast, producing a large amount of gas in a matter of milliseconds.

The sodium metal that is produced is highly reactive and can react with water to produce sodium hydroxide ($NaOH$) and hydrogen gas (H_2). To prevent this from happening, the airbag module also contains other chemicals that neutralize the sodium metal and any other potentially reactive products.

In conclusion, the chemical reaction that generates the gas in an airbag is a complex and carefully controlled process. Sodium azide is an essential component of this reaction, reacting with iron oxide to produce the nitrogen gas that inflates the airbag and protects the occupants of the car in the event of a collision.

68

Why are raindrops round?

Raindrops are naturally spherical in shape because of the surface tension of water. Surface tension is the force that pulls water molecules together, causing them to form a sort of skin or membrane at the surface. This force is stronger than gravity on small objects, such as raindrops, so the drop will try to minimize its surface area to achieve the most stable shape possible, which is a sphere.

Another reason why raindrops are round is due to air resistance. As the raindrop falls through the air, it encounters resistance from the air molecules. The resistance is greater on the edges of the raindrop, causing it to smooth out and become more spherical.

However, not all raindrops are perfectly spherical. Larger raindrops can become distorted as they fall through the air, and sometimes they can even break apart into smaller drops. Additionally, wind and other factors can also cause raindrops to take on different shapes.

Overall, the round shape of raindrops is a result of the physics of surface tension and air resistance. This shape allows the raindrop to fall more efficiently through the air and reach the ground as quickly as possible.

69

Why is salt used on icy roads and antifreeze used in cars in cold weather?

Salt and antifreeze are two common substances used to combat the hazards of cold weather. While both help prevent accidents and damage caused by low temperatures, they work in different ways.

Salt is often used to melt ice on roads because it lowers the freezing point of water. When salt is applied to ice, it dissolves and creates a solution of salt water. The salt water has a lower freezing point than regular water, causing the ice to melt even at temperatures below 0°C. This makes the roads safer for vehicles and pedestrians.

Antifreeze, on the other hand, is used in cars to keep the engine from freezing. Antifreeze is a mixture of water and chemicals, such as ethylene glycol or propylene glycol, that lower the freezing point of the liquid. By lowering the freezing point of the liquid, antifreeze can prevent the water in the engine from freezing and causing damage. Antifreeze also raises the boiling point of the liquid, which helps the engine run more efficiently in hot weather.

It's important to note that both salt and antifreeze can have negative environmental impacts. Salt can be harmful to plants and wildlife and can also corrode metal and concrete over time. Antifreeze can be toxic if ingested by humans or animals and can also contaminate groundwater if not disposed of properly. It's important to use these substances responsibly and to properly dispose of any excess or waste.

70

In how many states of water at the same time?

Water can exist in three states at the same time: solid, liquid, and gas. This phenomenon is known as the triple point, which is the temperature and pressure at which the solid, liquid, and gas phases of a substance are in thermodynamic equilibrium. For water, the triple point occurs at a temperature of 0.01 degrees Celsius and a pressure of 611.73 pascals (or 0.006 atm). At this point, water can exist as ice, liquid water, and water vapor simultaneously.

It's worth noting that the triple point of water is used as a defining point for the Kelvin temperature scale, where it is assigned a value of exactly 273.16 K. This means that the Kelvin temperature scale is defined such that the difference between the triple point of water and absolute zero (the point at which all matter has zero thermal energy) is exactly 100 degrees.

71

What is the chemical structure of steel, and does it rust?

Steel is an alloy of iron and carbon, with other elements such as manganese, silicon, and sometimes other metals added in small amounts. The carbon content of steel typically ranges from 0.2% to 2.1% by weight, depending on the grade.

The chemical structure of steel varies depending on the specific type and grade, but it generally consists of a crystalline lattice structure of iron atoms with interstitial carbon atoms occupying the spaces between them. Other elements may also be present in the lattice structure, such as nickel, chromium, or molybdenum, which can help to improve the strength, corrosion resistance, and other properties of the steel.

Despite its strength and durability, steel is prone to rusting, which occurs when iron reacts with oxygen and water in the presence of an electrolyte, such as salt. This reaction produces iron oxide (rust), which can weaken the steel and cause it to corrode over time.

To prevent rusting, steel is often coated with other metals or materials, such as zinc (galvanizing), chromium (chromium plating), or paint. Additionally, stainless steel is a type of steel that contains a minimum of 10.5% chromium, which creates a protective oxide layer on the surface of the metal and makes it resistant to rust and corrosion.

72

Have you heard of liquid crystals?

liquid crystals are a fascinating and important class of materials that have unique optical and electrical properties. They are used in a wide range of technologies, from LCD screens and electronic displays to temperature sensors and drug delivery systems.

Liquid crystals are a type of material that exhibits properties of both liquids and solids. They flow like liquids, but also have some of the order and structure of solids. This is because the molecules in liquid crystals are arranged in an ordered way, with the long axes of the molecules aligned in a specific direction.

There are several types of liquid crystals, including nematic, smectic, cholesteric, and discotic. Nematic liquid crystals are the most common and are used in most liquid crystal displays. They are characterized by their rod-like molecules that align parallel to each other but are disordered in the perpendicular direction.

Liquid crystals have many unique properties that make them useful in various applications. For example, they have high optical anisotropy, which means that they transmit light differently depending on the direction of the light and the alignment of the molecules. This property makes them ideal for use in LCD screens, where the polarization of light can be manipulated to produce images.

Another unique property of liquid crystals is their sensitivity to changes in temperature, electric fields, and other environmental factors. This property has led to the development of liquid crystal thermometers, sensors, and other devices that can detect changes in the environment and respond accordingly.

Overall, liquid crystals are a fascinating and versatile class of materials that have had a significant impact on modern technology.

73

How do we get poisoned?

Poisoning occurs when a person is exposed to a harmful substance, whether by ingestion, inhalation, injection, or absorption through the skin or eyes. Poisoning can result in a range of symptoms, from mild to life-threatening.

There are many ways that people can be exposed to poisons. Some common causes of poisoning include:

1. Accidental ingestion: This can happen when a person, particularly a child, accidentally swallows a harmful substance, such as a cleaning product, medication, or poisonous plant.
2. Inhalation: Breathing in toxic substances such as carbon monoxide, lead, or pesticides can lead to poisoning.
3. Absorption: Poisonous substances can enter the body through the skin or eyes. This can occur when someone handles a toxic chemical without proper protective gear.
4. Injection: Injecting illicit drugs can expose users to a range of toxins, including adulterants and impurities in the drug itself.
5. Environmental exposure: People can be exposed to poisons in the environment, such as lead in old paint or contaminated soil and water.

Symptoms of poisoning can vary widely depending on the type and amount of poison involved, as well as the person's age, overall health, and other factors.

Common symptoms can include nausea, vomiting, dizziness, headaches, confusion, seizures, and difficulty breathing.

If you suspect someone has been poisoned, it is important to seek medical attention immediately. In some cases, prompt treatment can be lifesaving. Treatment for poisoning may include administering antidotes, providing supportive care to manage symptoms, or removing the poison from the body through procedures such as dialysis or stomach pumping.

74

How does ozone (O3) work?

Ozone (O3) is a naturally occurring gas that is composed of three oxygen atoms. It is created in the Earth's atmosphere when ultraviolet radiation from the sun interacts with molecular oxygen (O2). Ozone plays a crucial role in protecting the Earth from the sun's harmful UV rays.

In the Earth's upper atmosphere, ozone forms a layer that acts as a shield, absorbing most of the sun's high-frequency ultraviolet light. This helps prevent the light from reaching the Earth's surface, where it can cause skin cancer, cataracts, and other health problems in humans, as well as damage crops and other plant life.

Ozone is also used as a powerful oxidizing agent for various industrial and commercial applications, such as water purification and air sanitation. When ozone comes into contact with organic and inorganic substances, it reacts and breaks down the chemical bonds in those substances, rendering them harmless or less harmful.

However, at ground level, ozone can be harmful to human health, particularly for those with respiratory problems such as asthma. Ozone is a key component of smog, which can cause coughing, wheezing, and other respiratory symptoms when inhaled. It can also damage crops, forests, and other vegetation, and can contribute to climate change.

Overall, while ozone is a vital component of the Earth's atmosphere, it needs to be managed carefully to minimize its negative effects on human health and the environment.

75

What fuels spacecraft?

Spacecraft are fueled by a variety of propellants depending on the mission and the design of the spacecraft. The most common types of propellants used in spacecraft are chemical propellants and electric propulsion.

Chemical propulsion typically involves the use of liquid or solid fuels, which are burned with an oxidizer to produce thrust. The most commonly used chemical propellants are liquid hydrogen and liquid oxygen, which are used in rocket engines to provide the high thrust needed to escape Earth's gravity and reach space. Other common chemical propellants include hydrazine, which is used for attitude control and maneuvering thrusters, and solid rocket boosters, which use a solid fuel and oxidizer mixture.

Electric propulsion, on the other hand, uses electric or magnetic fields to accelerate a propellant to high speeds. This method of propulsion is less powerful than chemical propulsion but is much more efficient and can provide continuous thrust over a long period of time. Common types of electric propulsion include ion thrusters, which use an electric field to accelerate ions out of a nozzle, and Hall-effect thrusters, which use a magnetic field to accelerate ions.

In addition to these traditional propulsion systems, there are also experimental technologies being developed such as solar sails, which use the pressure of sunlight to provide propulsion, and nuclear propulsion, which would use a nuclear reactor to heat a propellant and create thrust.

Overall, the choice of propulsion system depends on the specific mission requirements, including the speed and direction of travel, the payload size and weight, and the distance to be traveled.

76

Does acid rain fall from the sky?

Acid rain is a form of precipitation that contains high levels of acidic compounds, such as sulfuric acid or nitric acid. These compounds are released into the atmosphere through various human activities, such as burning fossil fuels and industrial processes, as well as natural phenomena like volcanic eruptions. Once in the atmosphere, these acidic compounds react with other chemicals and eventually fall to the ground as acid rain.

Acid rain can have harmful effects on the environment, including damage to crops and forests, and can make lakes and streams too acidic for fish and other aquatic life. It can also damage buildings and monuments made of limestone and marble.

While acid rain can be harmful, it is important to note that not all rain is acidic. In fact, most rain is slightly acidic due to the natural presence of carbon dioxide in the atmosphere, which reacts with water to form carbonic acid. However, the acidity of rain can become more severe in areas with high levels of pollution.

Overall, acid rain is a complex environmental issue that requires continued efforts to reduce emissions of acidic compounds and mitigate its harmful effects on the environment.

77

The interesting properties of water!

Water is a ubiquitous substance that is essential for life on Earth. It is a simple molecule made up of two hydrogen atoms and one oxygen atom, but its unique properties make it an incredibly versatile and useful substance.

One of the most remarkable properties of water is its ability to exist in three states - solid, liquid, and gas - at temperatures and pressures commonly found on Earth. This makes it possible for water to exist in a wide range of environments, from the icy depths of Antarctica to the steamy jungles of the tropics.

Another important property of water is its high surface tension. This is the result of the hydrogen bonds between adjacent water molecules, which create a cohesive force that allows water to form droplets and other complex shapes. This property is particularly important in biological systems, where it allows for the formation of cell membranes and other structures.

Water also has a high heat capacity, which means it can absorb and store large amounts of heat energy without experiencing significant changes in temperature. This is why water is commonly used as a coolant in many industrial and mechanical applications.

In addition, water is an excellent solvent, which means it can dissolve a wide range of substances, including salts, sugars, and acids. This property makes it essential for many biological processes, such as the transport of nutrients and the removal of waste products.

Finally, water is an incredibly important molecule for life on Earth. It is involved in countless biological processes, from photosynthesis in plants to the regulation of body temperature in mammals. It is also an essential component of many foods and beverages, and is a critical resource for agriculture, industry, and human survival.

In conclusion, water is a fascinating substance with a wide range of unique and important properties. It plays an essential role in countless natural and human-made processes, and is a vital resource for life on Earth.

78

Who is Mendeleev?

Dmitri Mendeleev (1834-1907) was a Russian chemist who is best known for his work on the periodic table of elements. He was born in Siberia and studied at the University of St. Petersburg, where he became a professor of chemistry.

In 1869, Mendeleev published his famous periodic table, which arranged the elements in order of increasing atomic weight and grouped them according to their chemical properties. He left gaps in the table for elements that had not yet been discovered, and accurately predicted the properties of these elements based on their position in the table. This led to the discovery of several new elements that had been unknown at the time.

Mendeleev also contributed to the development of the gas laws, and conducted research on the properties of petroleum and its derivatives. He was a prolific writer and wrote several books on chemistry and other subjects.

Mendeleev is widely considered to be one of the most important chemists in history, and his periodic table remains one of the most fundamental tools in the field of chemistry today.

79

How is the rate of radioactive decay determined?

Radioactive decay is the process by which an unstable atomic nucleus emits radiation in the form of particles or energy. The rate of radioactive decay can be determined using several methods, including counting the number of decay events, measuring the amount of parent and daughter isotopes in a sample, and using half-life equations.

One method for determining the rate of radioactive decay is to count the number of decay events per unit of time. This can be done using a device called a Geiger counter, which measures the number of ionizing particles or photons emitted by the sample. The rate of decay is proportional to the number of radioactive atoms in the sample, and can be expressed as the decay constant or activity.

Another method for determining the rate of decay is to measure the amount of parent and daughter isotopes in a sample. This can be done using techniques such as radiometric dating, which relies on the fact that the ratio of parent to daughter isotopes changes over time as the parent isotopes decay. By measuring the ratio of parent to daughter isotopes in a sample, scientists can determine the age of the sample and the rate at which the parent isotopes decay.

A third method for determining the rate of radioactive decay is to use half-life equations. The half-life of a radioactive isotope is the amount of time it takes for half of the original sample to decay. By measuring the amount of parent and daughter isotopes at different times, scientists can use half-life equations to calculate the rate of decay.

Overall, the rate of radioactive decay is an important concept in nuclear physics and is used in a wide range of applications, from dating ancient artifacts to determining the safety of nuclear reactors.

80

What is in a battery?

A battery is a device that converts chemical energy into electrical energy. It is made up of three basic components: two electrodes, an electrolyte, and a separator.

The two electrodes are usually made of different metals or metal compounds. One electrode, called the cathode, has a positive charge, while the other, called the anode, has a negative charge. When the battery is in use, electrons flow from the anode to the cathode through an external circuit, creating an electrical current.

The electrolyte is a substance that allows the flow of ions between the two electrodes, while preventing the flow of electrons directly between them. In most batteries, the electrolyte is a liquid or gel substance containing salts, acids, or bases. The ions in the electrolyte allow the chemical reactions at the electrodes to occur, which generate the electrical energy.

The separator is a thin layer of material that physically separates the two electrodes, while allowing the flow of ions through the electrolyte. The separator is usually made of a porous material that is resistant to chemical reactions and electrical current flow.

Overall, the chemical reactions that occur within the battery determine the amount of electrical energy it can produce. Different types of batteries use different chemical reactions, which can affect factors such as battery life, capacity, and rechargeability.

81

Why is the minimum temperature -273°C?

The minimum temperature in the universe is known as absolute zero, which is approximately -273.15°C or -459.67°F. This temperature is known to be the coldest possible temperature in the universe.

The reason behind this is based on the behavior of particles, specifically the particles that make up matter, such as atoms and molecules. These particles are always in motion, even at low temperatures, and their motion corresponds to a certain amount of energy. Absolute zero is the point at which the particles have the lowest possible energy, meaning they would have no motion or vibration.

According to the third law of thermodynamics, it is impossible to reach absolute zero by any means in a finite number of steps. In other words, the closer a system gets to absolute zero, the more difficult it is to reduce its temperature further. As a result, achieving a temperature of absolute zero is considered impossible in practice.

However, scientists can get very close to absolute zero by using a variety of techniques, such as laser cooling or magnetic cooling, which can reduce the temperature of certain materials to just a few billionths of a degree above absolute zero. This allows scientists to study the behavior of particles at extremely low temperatures, which can lead to new discoveries and technologies.

82

What happens when a chemist falls in love?

The chemistry of love is a fascinating and complex topic that has been studied by scientists for years. When a chemist falls in love, they are experiencing a variety of chemical reactions in their brain and body.

One of the primary chemicals involved in love is dopamine, a neurotransmitter that is associated with pleasure and reward. When a person is in love, their brain releases dopamine, which can create feelings of happiness, euphoria, and excitement.

Another important chemical involved in love is oxytocin, which is often referred to as the "cuddle hormone." Oxytocin is released during physical touch, such as hugging or kissing, and can promote bonding and trust between individuals.

The hormone adrenaline also plays a role in the chemistry of love. Adrenaline is responsible for the "fight or flight" response and can create feelings of nervousness, anxiety, and excitement.

Finally, there is the hormone serotonin, which is associated with feelings of well-being and happiness. When two people are in love, serotonin levels in their brains can increase, leading to a greater sense of contentment and satisfaction.

When all of these chemicals are working together, it can create a powerful and intense feeling of love. While the chemistry of love is not fully understood, scientists continue to study it in order to gain a better understanding of this complex and fascinating phenomenon.

83

What are nuclear weapons?

Nuclear weapons are powerful devices that use nuclear reactions to release an immense amount of energy in the form of an explosion. These weapons use either nuclear fission or a combination of nuclear fission and fusion to create an explosive force that can cause widespread destruction and devastation.

Nuclear fission is the process in which the nucleus of an atom is split into smaller fragments by bombarding it with neutrons. This process releases a large amount of energy and more neutrons, which can then initiate a chain reaction, leading to a massive release of energy in the form of an explosion.

Nuclear fusion is the process of combining two or more atomic nuclei into a single, heavier nucleus. This process also releases a tremendous amount of energy, as the combined mass of the fused nuclei is less than the mass of the individual nuclei.

Both nuclear fission and fusion can be used in the construction of nuclear weapons. In nuclear fission weapons, a critical mass of fissile material, such as uranium or plutonium, is assembled to create a chain reaction that releases an explosive amount of energy. In fusion weapons, a small fission explosion is used to compress and heat a fusion fuel, which then undergoes a fusion reaction, releasing a much larger amount of energy.

The use of nuclear weapons has been a topic of controversy and debate since their creation, as they have the potential to cause immense harm to human life and the environment. Many countries possess nuclear weapons as a form of deterrence, but the use of such weapons is strictly regulated by international treaties and agreements. The detonation of nuclear weapons can cause widespread destruction and loss of life, as well as long-term health effects due to radiation exposure.

84

Do metals get tired?

Metals can indeed experience "fatigue," which is a term used to describe the weakening of a material due to repeated or cyclic loading. This phenomenon is commonly seen in metal components that undergo repeated stresses, such as airplane wings or car parts.

Fatigue occurs because the repeated loading causes microscopic cracks to form in the metal. Over time, these cracks can grow and weaken the material, eventually leading to failure. The rate at which fatigue occurs depends on several factors, including the type of metal, the applied stress, and the number of cycles of loading.

However, it's important to note that not all metals experience fatigue to the same degree. Some materials, such as aluminum alloys, are more susceptible to fatigue than others, such as titanium alloys. Additionally, careful engineering and design can help minimize the effects of fatigue, such as by incorporating stress-relieving features or using thicker or stronger materials.

In summary, while metals can experience fatigue, the degree to which it occurs depends on several factors and can be mitigated through careful design and engineering.

85

Chemical weapons!

Chemical weapons are weapons that use chemicals to harm or kill living organisms, including humans. They are typically categorized as weapons of mass destruction, along with biological and nuclear weapons. Chemical weapons can be dispersed as a gas, liquid, or solid, and their effects can range from temporary incapacitation to death.

There are several types of chemical weapons, including nerve agents, blister agents, choking agents, blood agents, and riot control agents. Nerve agents, such as sarin and VX, are the most deadly and can cause death within minutes of exposure. Blister agents, like mustard gas, cause blistering of the skin and respiratory system, and can lead to long-term health effects. Choking agents, such as chlorine gas, cause damage to the lungs and can lead to suffocation. Blood agents, like hydrogen cyanide, interfere with the body's ability to use oxygen, causing death by suffocation. Riot control agents, such as tear gas, are intended to cause temporary incapacitation rather than death.

Chemical weapons have been used throughout history, with some of the most notorious instances occurring during World War I and the Iran-Iraq War. The use of chemical weapons is prohibited under international law, specifically the Chemical Weapons Convention, which has been signed by over 190 countries.

The production, stockpiling, and use of chemical weapons are considered war crimes and crimes against humanity. Despite the prohibition, chemical weapons continue to be a threat, with reports of their use in conflicts in Syria and elsewhere. The international community remains committed to preventing the use of chemical weapons and holding those responsible accountable for their actions.

86

Have you ever seen hormone-free food?

Hormone-free food has become a buzzword in recent years, as more and more consumers become concerned about the use of hormones in food production. The truth is, however, that there is no such thing as hormone-free food.

Hormones are naturally occurring substances found in all living organisms, including plants and animals. They play important roles in regulating various physiological functions such as growth and development, metabolism, and reproductive processes.

In agriculture, hormones are sometimes used to increase the growth rate and feed efficiency of animals, particularly in beef and dairy cattle. The most commonly used hormones are estrogen, progesterone, and testosterone, which are administered to animals in the form of pellets or injections.

While some consumers are concerned about the potential health effects of consuming meat or dairy products from animals that have been treated with hormones, the scientific consensus is that the levels of hormones in these products are safe for human consumption. In fact, the levels of hormones found in meat and dairy products are typically much lower than the levels naturally occurring in human bodies.

Nonetheless, some farmers and food producers choose to market their products as "hormone-free" in order to appeal to consumers who are concerned about hormone use in food production. However, it is important to note that these products are not necessarily any safer or healthier than products from animals that have been treated with hormones.

In summary, all food, both plant and animal-based, contains naturally occurring hormones. While some farmers choose to use additional hormones to increase the growth rate and feed efficiency of their animals, the levels of hormones found in meat and dairy products are generally safe for human consumption. As such, the label "hormone-free" does not necessarily indicate a safer or healthier food product.

87

Are you curious about the minerals in The World?

Minerals are one of the most valuable resources on the planet, with countries around the world producing various types of minerals that are essential for everyday life. From the production of smartphones to the manufacturing of airplanes, minerals play a crucial role in various industries. In this article, we will explore the most productive minerals of countries around the world.

1. China - Coal

China is the world's largest producer and consumer of coal, producing over 3.5 billion tons of coal in 2020 alone. Coal is a major source of energy in China, with the majority of the country's electricity generated from coal-fired power plants. China's coal industry has faced criticism for its impact on air quality and climate change, but the country continues to rely heavily on this mineral for its energy needs.

1. Australia - Iron Ore

Australia is the world's largest producer of iron ore, with over 900 million metric tons of iron ore produced in 2020. Iron ore is used to make steel, which is essential for the construction of buildings, bridges, and other infrastructure. Australia's iron ore exports are a crucial part of the country's economy, accounting for over 20% of its total exports.

1. Chile - Copper

Chile is the world's largest producer of copper, with over 5.7 million metric tons of copper produced in 2020. Copper is an essential mineral used in the manufacturing of electronics, electrical wiring, and plumbing. The country's copper industry has been a major contributor to its economy, accounting for over 10% of its GDP.

· **Russia - Nickel**

Russia is the world's largest producer of nickel, with over 270,000 metric tons of nickel produced in 2020. Nickel is an essential mineral used in the manufacturing of stainless steel, batteries, and electronic devices. Russia's nickel industry is a significant contributor to its economy, with the country exporting the majority of its nickel to countries around the world.

· **Canada - Potash**

Canada is the world's largest producer of potash, with over 12 million metric tons of potash produced in 2020. Potash is a mineral used in the production of fertilizer, essential for agricultural production. Canada's potash industry is a crucial part of the country's economy, with the majority of its potash exported to countries around the world.

· **Brazil - Bauxite**

Brazil is the world's third-largest producer of bauxite, with over 36 million metric tons of bauxite produced in 2020. Bauxite is an essential mineral used in the production of aluminum, which is used in various industries, including construction, transportation, and packaging. Brazil's bauxite industry is a significant contributor to its economy, with the country exporting the majority of its bauxite to countries around the world.

· **South Africa – Gold**

South Africa is one of the world's largest producers of gold, with over 118 metric tons of gold produced in 2020. Gold is an essential mineral used in the production of jewelry, electronics, and various industrial applications. South Africa's gold industry is a crucial part of its economy, with the country exporting the majority of its gold to countries around the world.

· **Indonesia – Tin**

Indonesia is the world's largest producer of tin, with over 70,000 metric tons of tin produced in 2020. Tin is an essential mineral used in the production of solder, which is used in various electronics and electrical applications. Indonesia's tin industry is a significant contributor to its economy, with the country exporting the majority of its tin to countries around the world.

· Australia:

Iron ore Australia is the largest producer of iron ore in the world, accounting for around 37% of total global production. The majority of Australia's iron ore exports go to China, which is the world's largest consumer of iron ore. Iron ore is used primarily in the production of steel, making it a crucial mineral for many industries around the world.

Russia:

Platinum group metals Russia is the largest producer of platinum group metals (PGMs) in the world, accounting for around 40% of total global production. PGMs include platinum, palladium, rhodium, and other precious metals, which are used primarily in the production of catalytic converters for automobiles, as well as in the jewelry and electronics industries.

Here are the top ten countries with the highest percentage of mineral reserves:

1. Russia - 30% of the world's mineral reserves
2. Democratic Republic of Congo - 10% of the world's mineral reserves
3. China - 9% of the world's mineral reserves
4. Australia - 6% of the world's mineral reserves
5. India - 6% of the world's mineral reserves
6. USA - 5% of the world's mineral reserves
7. Brazil - 5% of the world's mineral reserves
8. Canada - 4% of the world's mineral reserves
9. South Africa - 3% of the world's mineral reserves
10. Chile - 3% of the world's mineral reserves

It is worth noting that mineral reserve estimates can vary depending on the methodology used and the timing of the estimate.

88

What is borosilicate glass?

Borosilicate glass is a type of glass that is known for its exceptional durability and thermal shock resistance. It is made from a special blend of silica and boron trioxide, which gives it unique properties that make it highly desirable in many applications.

Borosilicate glass was first developed in the late 19th century by the German glassmaker Otto Schott. He was looking for a way to create a glass that could withstand rapid temperature changes without cracking or breaking, and found that adding boron oxide to the glass formula made it much stronger and more resistant to thermal shock.

The most common type of borosilicate glass used today is known as "Type I" glass. This glass is highly resistant to thermal shock and can withstand temperatures up to 450°C without breaking. It is also highly resistant to chemical corrosion and is therefore used in many laboratory and scientific applications.

One of the most well-known brands of borosilicate glass is Pyrex. Pyrex was first introduced in 1915 and quickly became a popular choice for kitchenware, laboratory glassware, and other applications that required a strong, durable glass. Pyrex is still widely used today and is known for its ability to withstand high temperatures and resist thermal shock.

Borosilicate glass is also commonly used in the manufacture of high-quality optical lenses and mirrors. Its low coefficient of thermal expansion makes it an excellent choice for precision optics, as it can maintain its shape and refractive index even when subjected to extreme temperature changes.

Overall, borosilicate glass is an incredibly useful material with a wide range of applications in science, industry, and everyday life. Its unique properties make it highly desirable for many different purposes, and it is likely to continue to be an important material for many years to come.

89

What is the structure of detergents?

Detergents are a class of compounds used for cleaning and are found in various household cleaning products, such as laundry detergents, dishwashing detergents, and bathroom cleaners. The chemistry of detergents is complex, but understanding the basic structure can provide insight into how they work.

Detergents are composed of two main components: a hydrophilic (water-loving) head and a hydrophobic (water-hating) tail. The hydrophilic head is typically composed of a polar group, such as a sulfate or a carboxylate group, while the hydrophobic tail is composed of a long hydrocarbon chain. This combination of polar and nonpolar groups allows detergents to interact with both water and oils or fats, making them effective at removing dirt and stains.

The hydrophilic head of a detergent molecule is attracted to water and can form hydrogen bonds with water molecules, allowing the detergent to dissolve in water. The hydrophobic tail, on the other hand, is repelled by water and is attracted to oily or fatty substances. When a detergent is added to water, the hydrophobic tails of the detergent molecules cluster together and form micelles, which are small, spherical structures with the hydrophobic tails pointing inward and the hydrophilic heads pointing outward.

When a detergent is used to clean, the hydrophobic tails of the detergent molecules interact with oils, grease, and dirt, while the hydrophilic heads interact with water. This allows the dirt and oils to be lifted away from the surface being cleaned and carried away in the water. Additionally, detergents can help to emulsify fats and oils, breaking them down into smaller droplets that are more easily removed.

Detergents can be classified into two main categories: anionic and nonionic detergents. Anionic detergents have a negatively charged hydrophilic head and are commonly used in laundry detergents and household cleaners. Nonionic detergents do not have a charged head and are typically used in industrial applications, such as in the manufacturing of paper and textiles.

Overall, detergents are an essential part of modern cleaning products. Their unique structure allows them to effectively remove dirt and stains from a variety of surfaces, making our daily lives cleaner and more hygienic.

90

What is a touchstone?

A touchstone is a small, dark, and fine-grained rock used to test the quality and purity of precious metals such as gold and silver. It is typically made of basalt or jasper and has been used since ancient times as a way to determine the authenticity of metal objects.

To use a touchstone, a small sample of the metal in question is rubbed against the surface of the touchstone, leaving a visible mark or streak. This mark is then compared to marks made by known samples of gold or silver to determine the purity of the metal. The touchstone allows for a simple and reliable method of assessing the quality of precious metals without the need for expensive equipment or chemical tests.

The name "touchstone" comes from the idea that the stone is a way to "touch" or test the quality of the metal, and to separate genuine gold or silver from counterfeit or adulterated versions. Touchstones have been used by goldsmiths, jewelers, and other metalworkers for centuries, and continue to be used in modern times as a reliable and low-tech method of testing precious metals.

91

Should we extract with cyanide or without it?

Cyanide is a highly toxic chemical compound that is known to be dangerous to human health and the environment. Despite this, it is widely used in the mining industry for extracting gold and silver from ores. Cyanide is particularly effective in separating these precious metals from the surrounding rock, but its use has also been associated with a number of environmental disasters, such as spills and leaks that have resulted in the contamination of rivers and other water sources.

The use of cyanide in mining has sparked significant controversy over the years, with some arguing that it is a necessary evil for extracting precious metals, while others argue that the risks associated with its use far outweigh the benefits. In response to these concerns, a number of countries have either banned or severely restricted the use of cyanide in mining, while others continue to use it under strict regulations and oversight.

There are a number of alternative methods for extracting precious metals from ores that do not involve the use of cyanide. These include methods such as gravity separation, flotation, and leaching with alternative reagents, such as thiosulfate or bromide. While these methods may not be as effective as cyanide extraction in all cases, they are generally considered to be safer and more environmentally friendly.

Cyanide gold mining is still a backward technology, and it is the activity of first crumbling rocks and soils containing micron-sized gold in very small sizes and then separating the gold in this material by leaching with cyanide. In such rocks containing gold, besides silver; Carcinogenic heavy metals such as arsenic, mercury and lead are also present at high levels. However, while these heavy metals are harmless because they are found in the natural environment in salt form, they become elements in cyanide leaching and are mobilized. Now the sleeping giant has awakened and it is inevitable that the rock material mud left after the gold and silver is taken will be a very dangerous waste. According to the EIA report containing one kilogram of rock material in the Bergama-Ovacık gold mine sample; 1.2 grams of Arsenic, 0.8 grams of Mercury, 0.8 grams of Lead etc. are at high levels.

In the cyanide leaching process, gold rock is crushed and ground. The ground ore is piled on a plane made almost impermeable (not completely impermeable) with a clay or polyethylene cover, as hills with a height of 3 to 70 meters, and a 0.05% NaCN solution is sprayed on these heaps. With this old method, which is still applied in many places in the world, 0.5-2.0 kg per 1 ton of ore. NaCN is consumed. The sodium cyanide solution drains to the bottom, extracting the gold in the ore heap. The duration of the leaching process can vary from a few weeks to several months.

Crude gold is obtained by electrolysis from the concentrated gold solution obtained as a result of stripping the activated carbon using a small amount of liquid. When heap leaching, the washed down ore hills are left where they are, while the fine sludge process wastes of cyanide leaching in stirred tanks are sent to the cyanide dam as they are or after cyanide oxidation.

If we list the environmental effects and weak points that may occur during or after this process;

1. Cracks, tears or punctures in clay pads or geotextiles.
2. Pipe bursts in cyanide solution or sludge transmission line.
3. The leachate reaches the underground by washing the waste piles stored in the open with rain.
4. Dissemination of cyanide solutions and sludges to the environment as a

result of the flooding of the tailings dam or the rupture of the bank.

5. The failure of the dam or the formation of cracks due to disasters such as earthquakes

6. Accidents (including traffic) that may occur during the transport, storage or transmission of cyanide

7. Environmental load to be created by nitrogen compounds from ammonia to nitrate resulting from the decomposition of HCN gas or cyanide carried into the atmosphere from the leach stage or tailings dam

8. Normally stable lead, cadmium, mercury etc. It is the mobilization of heavy metals such as cyanide as a result of grinding and cyanide analysis and gaining biological accessibility.

In the light of all these effects and possibilities, it is clear that cyanide will cause harm to our country and lands.

Ultimately, the decision to extract with or without cyanide will depend on a number of factors, including the type of ore being mined, the available technology and resources, and the regulations and restrictions in place in the country or region where the mining is taking place. It is important for mining companies and governments to carefully consider the risks and benefits of each approach, and to prioritize the safety of human health and the environment in their decision-making processes.

92

Do heavy metals settle in our stomachs?

Heavy metals are a group of elements that are naturally occurring and have a high density and atomic weight. These metals can be toxic in certain amounts and may accumulate in the body over time, leading to adverse health effects. While heavy metals can be found in a variety of sources, including air and water pollution, some people may wonder if heavy metals settle in our stomachs.

The short answer is yes, heavy metals can settle in our stomachs. When we ingest food and liquids that are contaminated with heavy metals, they can be absorbed by the stomach lining and enter the bloodstream. Once in the bloodstream, these metals can be transported throughout the body and accumulate in organs such as the liver, kidneys, and bones.

The accumulation of heavy metals in the body can lead to a variety of health problems, including neurological damage, liver and kidney damage, and cancer. Some common heavy metals that can be found in food and drink include lead, mercury, cadmium, and arsenic.

To minimize the risk of heavy metal exposure, it is important to be aware of potential sources of contamination and take steps to avoid them. This may include choosing products that have been tested for heavy metal contamination, avoiding contaminated water sources, and being cautious about consuming certain types of fish that may contain high levels of mercury.

In conclusion, heavy metals can settle in our stomachs and accumulate in the body, leading to serious health problems. It is important to be aware of potential sources of heavy metal contamination and take steps to minimize exposure.

93

How much radiation is harmful?

Radiation is all around us, and it's natural. The sun emits radiation, as do rocks and even some foods. Radiation is also produced by human-made sources such as X-rays and nuclear power plants. While low levels of radiation exposure are generally considered safe, exposure to high levels of radiation can be harmful to human health.

The unit of measurement for radiation is the Sievert (Sv), which measures the amount of radiation absorbed by human tissue. Exposure to just one Sievert of radiation can cause radiation sickness, which can lead to nausea, vomiting, and even death. Long-term exposure to radiation can also lead to an increased risk of cancer and other health problems.

The International Commission on Radiological Protection (ICRP) has set limits on the amount of radiation exposure that people can safely receive. The limit for radiation workers is 20 millisieverts (mSv) per year, while the limit for the general public is 1 mSv per year. However, these limits are subject to ongoing review and may change based on new research.

It's important to note that the risk of harm from radiation exposure depends on the type of radiation, the duration of exposure, and the individual's age and overall health. Children and fetuses are more susceptible to the harmful effects of radiation than adults, and people with weakened immune systems may also be at higher risk.

In summary, the amount of radiation that is harmful depends on a variety of factors, including the type of radiation, the duration of exposure, and the individual's age and health status. It's important to take precautions to minimize exposure to radiation whenever possible, and to follow recommended safety guidelines to protect yourself and others from harm.

94

What if everything was black and white?

What if everything was black and white?

The world around us is full of vibrant colors that add to the beauty and diversity of life. But what if everything was black and white? What would it be like to live in a monochromatic world where color did not exist?

First, it's important to understand that colors are created by the way light is absorbed and reflected by objects. In a black and white world, this would mean that all objects would appear in shades of gray, from pure white to pure black. This would include the clothes we wear, the buildings we live in, and even our skin.

Without color, it would be difficult to differentiate between objects based on their hue or saturation. This would make everyday tasks like picking out clothes or finding items in a cluttered room much more challenging. We would rely more heavily on texture and shape to identify objects and distinguish them from one another.

Another significant impact would be on art and media. Movies, television, and photography would all be monochromatic, with no color to add visual interest or emotional depth. Artists would need to rely solely on shading and contrast to create depth and perspective in their work.

Additionally, colors play an important role in signaling and conveying information. For example, traffic lights use red, yellow, and green to signal when to stop, slow down, and go. Without color, these signals would need to be replaced with alternative methods to convey the same information.

While a black and white world would be drastically different from the one we know, it's important to note that it wouldn't necessarily be worse. Humans are incredibly adaptable, and we would likely find new ways to navigate and appreciate a world without color. However, it's safe to say that the vibrancy and beauty of the world we live in would be sorely missed.

95

The laughing gas!

Nitrous oxide, commonly known as "laughing gas," is a chemical compound with the formula N_2O. It is a colorless, odorless, and non-flammable gas with a slightly sweet taste. Nitrous oxide is a powerful sedative and analgesic agent used in medicine and dentistry for pain relief and anxiety reduction during procedures.

In addition to its medical applications, nitrous oxide is also used as a recreational drug due to its euphoric and hallucinogenic effects. When inhaled, it produces a rapid onset of feelings of euphoria, relaxation, and dissociation. However, it can also cause side effects such as dizziness, nausea, and headache.

Nitrous oxide is produced naturally in the atmosphere by microbial processes in soil and water. It is also produced as a byproduct of burning fossil fuels and from agricultural activities such as fertilizer use and livestock manure.

Nitrous oxide is a greenhouse gas and a contributor to climate change. It has a global warming potential 298 times greater than carbon dioxide over a 100-year timescale. As such, efforts are being made to reduce its emissions through better agricultural practices and the use of alternative anesthetics and painkillers in medicine.

Despite its potential for abuse and environmental impacts, nitrous oxide remains an important tool in medicine and dentistry, providing a safe and effective means of pain relief and anxiety reduction.

96

Is our planet as big as a house?

To explain this, let's show the vacuum inside the atom, which is the smallest piece of matter. Imagine a big stadium. An insect in the middle. Here is the example of the atom. The nucleus of the atom is as small as an insect in the huge stadium. The orbits in which the electrons travel are also the walls of the stadium. Reduce the beetle in the middle of the stadium a hundred times. This is the size of the electrons. Just as such a stadium is empty, secluded and strange, so are atoms, and therefore our world. So, what is it that makes the objects around us seem full to us? It is the speed of electrons from one thousand kilometers per second to 15 thousand kilometers per second. If Mevla had not wisely rotated electrons around the nucleus, like a madman, this void within atoms would not have existed. And as a result, our big world would shrink as a house. In such a material structure, a teaspoon of water would weigh 1 billion tons.

In his experiment, Rutherford (1911) found that only one-thousandth of alpha particles sent from a thin sheet of metal (gold) go astray, most of which pass freely directly through the metal. Gold was also made of atoms. So the atoms were hollow.

What shows that the building blocks of matter, which we call atoms, are sparse and even empty, are these interesting events in space. Cosmic rays, which are streams of subatomic particles, enter from one end of a huge planet in front of them. They leave the other side of the planet without touching anything in their path.

If we were to fill an atom with its nucleus, which we liken to an insect in the middle of the stadium, 1015 nuclei would be required. We can briefly express the size of 1015 as one billion million. In other words: you will find a thousand times a million, you will get a thousand times this number again, you will get a thousand times the number again, and finally you will get the number 1015 again.

Of course, we consider this gap in matter from a material point of view. Otherwise, there are many secrets and wisdoms in matter. Even physicists investigating the nature of space now encounter a sea of motion in which many ghostly particles circulate in space.

97

Aluminum's dream!

At the beginning of the 20th century, a young student named Niels Bohr had a dream. He found himself standing in the center of the sun, surrounded by hot gases, while planets were orbiting around the sun with thin threads. Each planet sounded a whistle as it passed by Bohr. Then, the burning gases cooled and solidified, and the sun and planets moved away, and Bohr woke up.

This dream led him to think about a similarity between the solar system and atomic structure. This led to the creation of the "FIRST MODERN TABLE OF THE ATOM," which consisted of a nucleus surrounded by electrons. Thus, modern atomic theory began with a dream.

Another famous chemist, Kekule, had a similar experience. He wrote, "I turned my chair towards the fireplace and fell into a doze. Again the atoms were gamboling before my eyes. This time the smaller groups kept modestly in the background. My mental eye, rendered more acute by the repeated visions of the kind, could now distinguish larger structures of manifold conformation: long rows, sometimes more closely fitted together; all twining and twisting in snake-like motion. But look! What was that? One of the snakes had seized hold of its own tail, and the form whirled mockingly before my eyes. As if by a flash of lightning, I awoke."

In his dream, Kekule saw a snake biting its own tail, which led to his discovery of the ring shape of benzene (usually shown as a hexagon) and the "Closed Chain" or "Ring" theory in organic chemistry that highlights the importance of molecular structure.

Both Bohr and Kekule had dreams that inspired their scientific discoveries. Dreams have a powerful impact on human imagination, and sometimes they can lead to scientific breakthroughs.

98

How much pressure does the atmosphere exert on our bodies?

The atmosphere is the layer of gases that surrounds the Earth, extending about 6,000 km (3,700 miles) from the surface. It is composed of mainly nitrogen (78%) and oxygen (21%), with trace amounts of other gases such as argon, carbon dioxide, and neon. Due to the weight of this gas, the atmosphere exerts a pressure on the surface of the Earth and everything on it, including our bodies.

The pressure exerted by the atmosphere is known as atmospheric pressure, and it varies with altitude. At sea level, the average atmospheric pressure is about 101.3 kilopascals (kPa), which is equivalent to 14.7 pounds per square inch (psi). This means that the atmosphere is pushing down on every square inch of our bodies with a force of 14.7 pounds.

As we go higher in altitude, the atmospheric pressure decreases, and so does the amount of oxygen available for us to breathe. This is why climbers and pilots often carry oxygen tanks when ascending to high altitudes.

The human body is able to withstand changes in atmospheric pressure to a certain extent, as our bodies are designed to maintain a balance of pressure both inside and outside. However, rapid changes in pressure, such as those experienced during scuba diving or flying in an unpressurized aircraft, can cause serious health problems if not managed properly.

In conclusion, the atmosphere exerts a constant pressure on our bodies, which is highest at sea level and decreases with altitude. While the human body can adapt to changes in pressure to a certain extent, rapid changes in pressure can be dangerous and require proper precautions to be taken.

99

Microbes help extract gold!

Gold is one of the most precious and valuable metals in the world, and its extraction has always been a challenging and costly process. But what if tiny microbes could help us extract gold from the earth? Recent research has shown that certain types of bacteria can help us do just that.

Traditionally, gold is extracted from ore by crushing it into a fine powder and then treating it with chemicals such as cyanide or mercury. These chemicals dissolve the gold, but they are also toxic and can cause serious environmental damage. Microbial gold extraction offers a more sustainable and environmentally friendly alternative.

The process involves using bacteria to dissolve the gold from the ore. The bacteria, which are found naturally in the soil, secrete a molecule called a ligand, which binds to the gold and dissolves it. The gold is then separated from the solution, and the bacteria can be reused.

There are many advantages to using microbes for gold extraction. For one, the process is much less harmful to the environment than traditional methods. Additionally, microbial gold extraction can be done at a lower cost than chemical methods, making it a more economical option.

Another advantage is that microbial gold extraction can be done in places where traditional mining methods are not possible. For example, it can be used to extract gold from low-grade ores or from tailings, which are the waste materials left over after the extraction of higher-grade ore.

Microbial gold extraction is still in the early stages of development, but it shows great promise as a sustainable and cost-effective method of extracting gold from the earth. With further research and development, it may become a key method of extracting this valuable metal in the future.

100

Soap in history!

Soap has been a part of human civilization for thousands of years, with evidence of soapmaking dating back to ancient Babylon around 2800 BCE. The earliest soaps were likely made from a combination of animal fat and wood ash, which produced a crude but effective cleaning agent.

The use of soap spread throughout the ancient world, with the Greeks and Romans using soap for personal hygiene and medicinal purposes. In medieval Europe, soap was primarily produced by monks and used by the wealthy as a luxury item.

The process of soapmaking evolved over time, with the development of new ingredients and techniques. In the 18th century, the discovery of the chemical composition of fats and oils led to the development of modern soapmaking methods. The addition of alkali, such as lye, to fats and oils created a chemical reaction known as saponification, which produced soap.

By the 19th century, soapmaking had become a major industry, with the invention of new cleaning agents and the development of mass production methods. Soap became more widely available and affordable for everyday use.

Today, soap is an essential part of daily life, used for personal hygiene, cleaning, and a wide range of other applications. The ingredients and formulations of soap have evolved to meet changing consumer preferences, with the addition of fragrances, moisturizers, and other additives.

While soap may seem like a simple and ubiquitous product, its history is a testament to the ingenuity and creativity of humans throughout the ages.

101

Patriotism!

"One of the driving forces of his work was the desire to elevate Poland's honor in a foreign country"

The discovery of radium broke new ground in medicine. The name of Marie Curie, who discovered radium, is associated with her husband, Pierre Curie, with whom she collaborated.

Marie Curie was a Polish born in Warsaw. One of the driving forces of his work was his desire to elevate Poland's honor in a foreign country. She never forgot Poland during the long years she lived in France. The fact that he was born at a time when his country was under Russian occupation always made him feel sad. One of the reasons he chose France for his scientific studies was that women were not allowed to study medicine in England at that time.

Madame Curie separated the peshblende ore used in the Radium experiments into its elements and named one of them after her hometown, POLONIUM. He also named one element radium.

Radium rays could pass through even the hardest objects. Thus, the radium became a starting point in the treatment of cancer.

Radium experiments were becoming very dangerous. During these experiments, Madame Curie often fell ill. Pierre Curie also burned his hands once.

The most interesting gift they received during their work was a ton of Peshblend items sent by the Austrian Emperor. Half a gram of radium could be obtained from one ton of Peshblende, provided that fifty tons of water and five to six tons of other chemicals are added.

In 1921, American women tied him a small salary. Madame Curie also spent the money to find a piece of radium at a hospital in Poland, which she loved so much.

The Curies both enjoyed a simple life. They were very humble. They lived so quietly that the world learned that Madame Curie was in a state of death two days before her death. Her husband, Pierre Curie, had previously died in an accident. Their worldly lives were separated, but their biographies were written together: Pierre and Marie Curie, (1859-1906, 1867-1934).

We should remind once again that Pierre and Marie Curie's love for their country is a source in their scholarly work.

It may make sense for those who have achieved success in scientific studies to say that science is "universal", "common property of humanity". But it doesn't make sense for those who always only gather the results to say this.

Humanity does not have a separate class called "Scientists". Every scientist is a part of his own country.

102

The unforgiving element!

Fluorine is a highly reactive element that is never found freely in nature. Its name is derived from the Greek word "fluere," which means "to flow," indicating its ability to react and flow easily with other elements. In fact, fluorine is so reactive that many scientists have lost their lives in attempts to obtain it.

Early efforts to extract fluorine were marked by accidents and injuries. Many scientists, including the Irish Academy of Sciences' Knox, French chemist Niklesse, Belgian researcher Layette, and famous chemists Gay-Lussac and Thenard, were victims of fluorine's destructive properties. English chemist Humphry Davy was also injured in his attempts to isolate the element.

It wasn't until 1886 that French chemist Henri Moissan successfully obtained free fluorine, but not without sacrificing one of his eyes in the process. Despite its dangers, today's scientists have found ways to use fluorine compounds for various applications, including the creation of freon gas for refrigeration.

The chemistry of fluorine has become a large and independent branch of inorganic chemistry, with many practical applications in industry and medicine. Although it is still a dangerous element to work with, the development of methods to control its reactivity has allowed scientists to harness its potential for the betterment of humanity.

103

How many types of water are there on Earth?

As it is known, there are three hydrogen isotopes (1H, 2D, 3T) in nature. Each of them combines with oxygen to form water, so we can talk about various waters (H_2O, D_2O, T_2O). It can also be in mixed waters with one atom of protium and one atom of deuterium, or one atom of deuterium and one atom of tritium in their molecules. Thus, the water types increase: HDO, HTO and DTO.

On the other hand, the oxygen contained in water is a mixture of three isotopes: Oxygen-16, oxygen-17, and oxygen-18. The first isotope is the most common.

Given this diversity of oxygen, 12 more possible waters are added to the list. When you take a glass of water from a lake or river, there is a possibility that you may have eighteen different types of water in your glass. The isotope component of water varies depending on its location. This is due to the large number of isotope substitution processes in nature. Different isotopes of hydrogen and oxygen constantly replace each other under various conditions.

D_2O, especially called heavy water, finds wide application in practice. In nuclear reactors, heavy water is used to slow down the neutrons that cause uranium fragmentation. Apart from that, scientists use various types of water in their studies in the field of isotope chemistry.

Water is one of the most essential and abundant substances on Earth. It is estimated that about 71% of the Earth's surface is covered with water. However, not all water is the same. There are different types of water on Earth, each with its own unique properties and characteristics. In this article, we will explore the different types of water found on Earth.

1. Saltwater: Saltwater is the most common type of water found on Earth, making up about 97.5% of all water on the planet. It is characterized by its high salt content, which makes it unsuitable for human consumption. Saltwater is found in oceans and seas all around the world.

2. Freshwater: Freshwater is the second most common type of water on Earth, making up only 2.5% of all water. It is characterized by its low salt content and is essential for human consumption and agriculture. Freshwater is found in rivers, lakes, and underground aquifers.

3. Groundwater: Groundwater is a type of freshwater that is found underground in the soil and rocks. It is an important source of water for many communities and industries around the world.

4. Surface water: Surface water is another type of freshwater that is found in rivers, lakes, and other bodies of water on the Earth's surface. It is an important source of water for irrigation and agriculture.

5. Glacial water: Glacial water is a type of freshwater that is found in glaciers and ice caps around the world. It is an important source of water for many communities and industries, particularly in regions where other sources of water are scarce.

6. Mineral water: Mineral water is a type of water that contains high levels of minerals such as calcium, magnesium, and potassium. It is believed to have health benefits and is often sold as a bottled drink.

7. Spring water: Spring water is a type of water that flows naturally to the surface from underground aquifers. It is often considered to be of high quality and is also sold as a bottled drink.

8. Distilled water: Distilled water is a type of water that has been purified through a distillation process, which removes all impurities and minerals. It is often used in laboratories and medical settings.

In conclusion, there are many different types of water on Earth, each with its own unique properties and characteristics. From the salty water of the oceans to the pure distilled water used in laboratories, water is a vital resource for all life on our planet.

www.ingramcontent.com/pod-product-compliance
Lightning Source LLC
Chambersburg PA
CBHW070538220526
45467CB00003B/984